INDUSTRIAL AIR
POLLUTION CONTROL

Kenneth Noll

Associate Professor of Civil Engineering
The University of Tennessee

Joseph Duncan

Assistant Director
Knox County Pollution Control

ann arbor science PUBLISHERS INC.
P.O. BOX 1425 • ANN ARBOR, MICHIGAN 48106

Preface

This book is concerned with a large variety of practical aspects of air pollution control for industry. The various sections provide information on the installation and operation of control systems in the broad categories of power generation, metallurgical processes, raw material processing, chemical and wood products, and industrial combustion. Within these broad categories, specific control techniques are explored for particles, sulfur oxides, nitrogen oxides, hydrocarbons, fluorides and odor. An introductory section provides an overview of the air pollution problem emanating from industrial operations.

The authors first presented this information at one of the air pollution control conferences held at Knoxville, Tennessee, chaired and managed by the authors.

It is hoped that this book will serve as a valuable guide to those engaged in air pollution control work by providing information on current air pollution control techniques for use as a foundation for more effective air pollution control.

Kenneth E. Noll
Joseph R. Duncan
April, 1973

CONTENTS

Section I

Introduction

1.
STATUS AND TRENDS IN INDUSTRIAL
AIR POLLUTION CONTROL

Arthur C. Stern

Professor, Department of Environmental Sciences and Engineering
University of North Carolina, Chapel Hill

There are two aspects of status and trends in industrial air pollution control. One is social and political; the other technological. Years ago, factories were supposed to belch smoke and fumes as a telltale that they were in operation and providing employment to people. In the common phraseology of the time, "when the factories stopped smoking" meant "the workers were idle." Factory owners insisted that the engravers and lithographers who made prints of their plants show dense plumes issuing from their chimneys. How else could they retain the respect of their peers? In similar fashion, the steam locomotive was invariably pictured with smoke bellowing from its stack to portray its power and its motion. These symbols were socially and politically expected and acceptable to all but a small minority who, in the 1880s, caused the birth of the smoke abatement movement in America. By 1907, there were enough people professionally concerned with smoke abatement in the United States to allow the formation in that year of the Smoke Prevention Association. Its membership was mainly railroad and municipal smoke inspectors.

Contrast this situation with that of today. Now no corporate executive in his right mind would allow a publicity picture of his plant to show even a wisp of smoke or fume issuing from its stacks. Any evidence of a whitish emission would be carefully captioned as innocuous steam even if, in fact, the steam masked a myriad of pollutants. The gospel today is that workers are repelled by a plant with belching stacks and are attracted to a plant with clean ones. The clean stack is considered the desirable norm; the dirty stack is labeled a malfunction of plant operation that management will correct as rapidly as possible. Pollution emissions, particularly if they are visible or odorous, are no longer socially or politically acceptable. The Smoke Prevention Association has changed its name to the Air Pollution Control Association. Its members are largely engineers, scientists, and other professional persons. One would have to look long and hard at its membership list to find a railroad or municipal smoke inspector.

FUTURE SOCIAL AND POLITICAL STATUS

The trend suggests that, in the future, our society and body politic will be no more satisfied with only our present concept of air pollution control than

1

they are now willing to settle for only our previous concept of smoke abatement. We already see the rash of public and private projects being halted, diverted, or delayed by environmentalists' action in the courts and in stockholder meetings. The trend of the future certainly seems to require the person concerned with air pollution control to be a member of a multidisciplinary site planning, land-use planning, natural-resource planning team. The social and political status of these teams and of their coordinators will be of the highest. They will have political visibility and political clout. Woe betide the plant or process designer who would convert an air pollution problem to one of water pollution, or who would not find a use for a waste product. The present Air Pollution Control Association—why, it will be a member of the as yet unformed Federation of Environmental Planning Associations of America.

TECHNICAL STATUS AND TREND

Technologically, industrial air pollution control takes three forms:
  1. Use of a tall stack to disseminate the pollution in such a manner that it will be diluted with air before it envelops a receptor.
  2. Use of a control device to decrease or eliminate pollution prior to emission.
  3. Change of process or of raw materials so as to decrease or eliminate pollution production.

The Tall Stack—The First Control Means

Historically, the tall stack was the first air pollution control means to be used by industry. It came into being because it provided the draft necessary to pull air through grates and products of combustion through furnaces and past products that required roasting, calcining, melting, smelting, evaporating, or drying. It released a mixture of combustion products and process effluents at the aboveground elevation determined necessary to produce the required draft. The stack height required to produce draft was small in comparison to what we today call a tall stack. However, there was no incentive to build taller stacks. In fact, there were several disincentives: one was the fact that a still taller stack produced too much draft, thereby requiring more elaborate dampers; another was that a taller stack was more expensive to build. In the era when these stacks were considered tall, buildings were low, so that there were three levels of pollution release to the atmosphere: refuse burned on the ground, chimney-pots at building roof-top level, and the tall (by comparison) industrial stacks of the day. All three levels were so close to the ground that their effluents quickly mixed into the same air stratum.

The changes that have taken place in our use of the tall stack are many. Today stacks are rarely designed as draft producers. We rely upon electric motor or turbine driven fans to provide the required process draft. Because of fans we can emit combustion products and process effluents through separate stacks or through a combined stack as we see fit. When separate stacks are used, they can be still further separated into low stacks or vents for innocuous effluents and tall stacks for objectionable effluents. The height of the tall stack is no longer determined by the need for draft but by the need to dilute the effluent on its path from the top of the stack back to ground level so that the diluted concentration is not objectionable. This can require very tall stacks. Today, some are as high as 1,000 feet.

At the same time that some industrial stacks have gotten taller, so have buildings. The ground level burning of refuse has almost disappeared as an important source, but we still have three significant levels of pollution release: the height of residential chimneys, the height of multistory buildings, and the height of industrial stacks. The taller the industrial stacks, the higher their plumes will rise due to buoyancy and momentum of release, and the effluents will start the dilution process very high above the ground. The air stratum in which this occurs can be so far removed from that into which the two lower level releases occur that they can be considered as polluting different air masses: the low level releases polluting close to the buildings involved; the high level releases polluting at groundlevel remote from the stacks involved.

The Future Status of the Tall Stacks

As we look into the future, we see the need for the tall stack receding. Nobody except stack construction companies really wants to build 1,000-foot stacks. They are visible proof of the inability of the designer to come up with a less costly solution. These less costly solutions will evolve as designers concentrate on process change, raw material change, and control device improvement . Originally, they may not be less costly than the tall stack. However, limitations on stack emission to prevent remote pollution problems and global atmospheric concentration build-up will require the use of these alternate solutions whether or not a tall stack is used. As long as the tall stack was allowed as the sole means for air pollution control, it could emerge as the least-cost solution. However, once additional solutions are required by regulation, the tall stack becomes a cost burden on the other solutions that these other solutions cannot afford to carry; as a result, the need for the very tall stack disappears.

CONTROL DEVICES—THE INTERIM SOLUTION

The equipment that we today recognize as control devices to decrease or eliminate pollution prior to emission was, in most cases, not originally intended for that purpose. The history of devices for the collection of particulate or gaseous matter from gas streams usually starts with their development and use to collect a useful product, such as flour or whiskey. There originally was very little incentive for an inventor to develop a device to prevent a particulate or gaseous waste from going to the atmosphere, but there was great incentive to improve the process of collecting, recovering, or recycling a valuable product or of removing from a valuable gaseous product an impurity that prevented its optimum use. Once devices for these purposes were perfected, it was but a short step to adapting them to the collection of waste which cost more to collect than it was worth as raw material or product improvement. However, what manufacturer would willingly burden his production cost by using a device that improved neither product quality nor quantity? The answer is that none would unless forced to do so by a governmental regulatory authority.

Today, governmental regulatory authority requires the degree of effluent control achievable by the use of control devices. The accretion of such governmental control requirements has been so rapid and so recent as to have given control equipment manufacturers a veritable bonanza of new business. Years ago each manufacturer touted the benefits of the one type of device he manufactured. As control equipment manufacture has become big business, this has largely disappeared. Today, each major manufacturer markets a broad spectrum of control device-types and selects from this array the device or combination of devices that best fits the customer's needs. Coupled with this is the increased ability of the customer's engineers to select the most applicable device or combination of devices. Because of this, my momentary temptation to discuss the available types of control devices and their application is readily suppressed.

THE FUTURE STATUS OF CONTROL DEVICES

At present, it is so easy for a designer to incorporate a control device into his process and so difficult to develop a new process that does not require the control device that it is small wonder designers follow the line of least resistance and select for their processes one of the array of excellent control devices now available. How long will this situation prevail? Most likely for a considerable period of time into the future because, as good as they are today, control devices will get still better as the new-found prosperity of the device manufacturer allows them a luxury they never had before: being able to support an adequate research and development effort. At the same time, industries, newly required by governmental regulatory authority to use control devices to an extent unheard of before, will try to make them less the "white elephants" they have been in the past. There will be increased pressure on plant designers and operators to find profitable uses for what have heretofore been considered wastes. Many will be successful in their quest.

PROCESS CHANGE—THE ULTIMATE SOLUTION

Now let's look at process and raw material change as the ultimate means for air pollution control. It is the ultimate means because a process or raw material that does not produce pollution does not have the problem of controlling it by control devices or tall stacks. One has only to read a history of technology, such as the five-volume set published by Oxford Press and edited by Charles Singer *et al.*, to recognize that industrial processes are evolutionary. If, instead of looking at the changes between 1971 and 1972 model automobiles, or even the 1962 and 1972 cars, one looks at the change in the means of transportation of the years 1772, 1872, and 1972, one sees evolutionary technology at work. We get the same picture if we look at making steel, alkali, acid, textiles, or any one of many other products. We would be foolish indeed to say that technology stops evolving in our generation. Quite the contrary should be the case. This generation has more and better trained scientists and engineers than any previous generation. This should increase the evolutionary pace.

Part and parcel of evolutionary technology has been the production of a unit of product with the emission to the atmosphere of less and less particulate and gaseous wastes; certainly, where the wastes have not decreased quantitatively, they have improved qualitatively by becoming less and less obnoxious. The noxious trades of one and two centuries ago are largely disappearing as noxious trades. Even the term "noxious trade" is in such present disuse that many of you may be reading it here for the first time. Even as the industrial operations that were anathema a few generations ago are scarcely recognizable today, so those of greatest concern today may require explanation to our grandchildren as to why we were so concerned about them today.

CONCLUSION

In conclusion, I note that I have succeeded in writing about industrial air pollution control without mentioning any air pollutant by name; discussing no specific law, regulation, or standard; describing no equipment or process; and naming almost no industries. This is as it should be. This introduction is followed by other sections, each intended to tell you about one of the subjects I so carefully avoided.

2.
AMBIENT AIR MONITORING FOR POINT SOURCES

Robert W. Garber, Gordon G. Park,
Auburn E. Owen, Jr., and Thomas L. Montgomery

Tennessee Valley Authority, Air Quality Branch
Muscle Shoals, Alabama

This paper outlines the general considerations used in establishing a point source ambient air monitoring program. In order to remain a general guide, no specific examples are used in this discussion; however, it does embody the approach used for over 20 years in the ambient air monitoring program of the Tennessee Valley Authority at the agency's coal-fired power plants.[1]

The principal objective of any ambient air monitoring program is to provide information which will aid in preserving the ambient air quality. In order that this goal is achieved in the most efficient and economical way, ambient air monitoring can be used to obtain data with respect to the nature of the atmospheric contaminants and to evaluate the effectiveness of air quality control programs. Although ambient monitoring is the responsibility of the state,[2] large industries will usually set up a monitoring program either to obtain information on air quality or to satisfy specific requirements of statutes or regulations. An example of these requirements is seen in some state regulations which provide alternative means of meeting ambient air criteria either through meeting emission standards or by demonstrating that emissions are not causing violations of ambient standards. In these cases, a source seeking to utilize the latter control strategy must establish an air monitoring program to demonstrate that ambient standards are being met.

In other states, there are no provisions for alternate control strategies and only emission standards are provided for.

PRELIMINARY CONSIDERATIONS

When establishing a new program or updating an existing one of ambient air monitoring for a point source, much information is required prior to actual implementation of the program. Preliminary studies should be carried out to determine (1) the method of dispersion, (2) the effects of other sources, and (3) the type of monitor to be used.

Method of Dispersion

One of the most fundamental requirements of a program of ambient air monitoring for a point source is the prediction of groundline concentrations

7

of released species. The dispersion characteristics of a plume are dependent upon the meteorological and topographical conditions, the method of release, and the physical condition of the effluent.[3] Calculations using appropriate models will yield maximum groundline concentrations and their distances from source. However, prior to performing dispersion calculations, certain parameters must be studied in detail.

The local meteorology is probably the most important parameter to be studied for the location of monitors with respect to a point source. The direction of prevailing winds and the local turbulent mixing directly affect the direction an effluent will be transported and the manner of dispersion in the atmosphere. Also, the necessary number of monitors will depend partially upon the local meteorology.

The topography of a region, indirectly through alteration of dispersion patterns, will influence the ambient air monitoring of point sources. This occurs through mechanical interaction of various structural features of the surface with the natural airflow causing local turbulence and mixing. Certain topographical features can result in direct plume visitation of an area of land due to its location with respect to the source, *i.e.,* cliffs and hills, located adjacent and above the source. Thus, topography must be considered in any monitoring plan.

Other Sources

The contribution of atmospheric pollutants by other sources must be taken into account to correctly ascertain the real contribution from the source of interest. Also, other sources may release substances which can interfere with the proposed method of measurement of the species of interest.

If an ambient air monitoring program is set up in advance of the startup of the source, a detailed history of the background levels of the species of interest can be obtained directly through measurement. Information of this type is necessary when it is required to determine the concentration con-tributed by a given source.

When an ambient air monitoring program is being set up for an existing source, background levels of species of interest must be obtained indirectly. Generally, such information is obtained via estimation based upon dispersion modeling.

Other sources can affect an air monitor by interference.[5] Agents released by another source which directly interfere with a given measurement must be known and action taken to adjust in the measurement. In most cases, some slight modification (filter, etc.) of the system will eliminate the interference.

Type of Monitor

There is a large variety of ambient air monitoring instruments available from commercial sources which utilize some chemical or physical property of the species to be measured. EPA recently promulgated requirements for ambient air monitors to be used implementing the Clean Air Act. These guidelines indicate that any technique which meets the sensitivity and specificity requirements will be approved (Table 1). Many techniques used for ambient air monitoring can be interfered with by other chemical species present in certain ambient environments. As these interferences can cause a high degree of error in a measurement, action must be taken to eliminate such interference prior to the analysis. Temperature has a pronounced effect upon the rate of chemical reactions and the operation of electronic circuitry, and so appropriate precautionary measures must be taken to protect equipment and results whenever the expected temperature variation exceeds the limits imposed by the system.[4]

Finally, the maintenance of the monitor should be considered when purchasing any instrumentation. In general, the less maintenance required for a given instrument, the lower the long-term cost of the monitoring program and the less will be the chance of maintenance-induced error.

In addition to automatic continuous air monitoring instrumentation, there are static-type monitors which are of use for air quality monitoring.[5] However, these systems cannot be directly quantitated in terms of concentration of species in air. Since they are quite inexpensive and require no field maintenance except field placement and removal at specified intervals, they can be used to evaluate long-term trends and general pollutant levels in a given region.

MONITORING PROGRAM

All air monitoring systems are for the general purpose of determining ambient air quality. These usually subdivide into (1) preoperational determination of background pollutant levels before plant construction and (2) postoperational documentation of ambient air quality after the plant goes into operation.

Preoperational

Background air quality information should be obtained and should span at least one year prior to plant operations to (1) document area air quality, (2) test monitoring equipment, (3) become familiar with evaluating air quality data, and (4) formulate plans based on field experience prior to actual plant operations.

Generally, monitoring for background information should consist of sampling for pollutants which are typical for the source in question. Criteria for locating monitors in the order of their relative importance are (1) one or more monitors in the direction of prevailing winds, (2) at least one monitor at the point of maximum plume impact (determined by appropriate dispersion model), (3) one or more monitors adjacent to nearby urban centers, and (4) one or more monitors near important agricultural crops or forests. If there appears to be significant interference from other sources in the area, it would be well to consider locating monitors in line with these sources and, ideally, in the direction of prevailing winds.

Postoperational

The collection of postoperational air pollution data may either be a continuation of monitoring begun in the plant preoperational period or may begin at a source already in operation. In either case, the general guidelines for monitoring should be (1) an assessment of ambient air quality to determine the new source contribution, (2) an opportunity to reposition monitors to accurately reflect maximum concentrations, (3) the comparison of air quality levels with state and local standards, and (4) the evaluation of the effectiveness of control equipment installed at the source.

In the case where the air monitoring program is to be conducted without the benefit of preoperational data, some modification may be required. Fundamental considerations for planning an air monitoring system, such as a study of local meteorology and topography, position of other sources, and land use patterns, should be followed. The basic difference in conducting this study will be to determine the background levels prior to operation of the new source. A special placement of air monitors and analysis of ambient data should differentiate between existing area sources.

Data Processing

It is well beyond the scope of this paper to review data reduction methods; however, no paper on monitoring would be complete without some comments on this topic. In many cases, the use of computer and automatic data processing techniques is absolutely essential to processing monitoring data due to the size of the program.[6]

There are some advantages with automatic data processing even with smaller operations. When the data are read directly into a computer via some form of digital output device (magnetic tape, punch tape, etc.), error generated in reading analog charts is eliminated and processing is accelerated. Also, random errors generated during repetitive hand calculations are eliminated. Careful consideration must be given to the methodology to be used for data processing during the preliminary preparations for an ambient air monitoring program.

CONCLUSIONS

The crucial role of ambient air monitoring in successful air quality maintenance is evident. In establishing an ambient air monitoring program, it should be emphasized that each situation requires a well-thought-out plan utilizing the detailed analysis of all significant parameters. For existing sources, effective modeling for the prediction of background levels is

Table 1. Ambient Air Monitor Requirements[1]

Specification	Sulfur Dioxide	Carbon Monoxide	Photochemical Oxidant (corrected for NO_2 and SO_2)
Range	0–2,620 μg/m^3 (0–1 ppm)	0–58 mg/m^3 (0–50 ppm)	0–880 μg/m^3 (0–0.5 ppm)
Minimum detectable sensitivity	26μg/m^3 (0.01 ppm)	0.6 mg/m^3 (0.01 ppm)	20 μg/m^3 (0.01 ppm)
Rise time, 90%	5 minutes	5 minutes	5 minutes
Fall time, 90%	5 minutes	5 minutes	5 minutes
Zero drift	±1% per day and ±2% per 3 days	±1% per day and ±2% per 3 days	±1% per day and ±2% per 3 days
Span drift	±1% per day and ±2% per 3 days	±1% per day and ±2% per 3 days	±1% per day and ±2% per 3 days
Precision	±2%	±4%	±4%
Operation period	3 days	3 days	3 days
Noise	±0.5% (full scale)	±0.5% (full scale)	±0.5% (full scale)
Interference equivalent	26μg/m^3 (0.01 ppm)	1.1 mg/m^3 (1 ppm)	20 μg/m^3 (0.01 ppm)
Operating temperature fluctuation	±5° C	±5° C	±5° C
Linearity	2% (full scale)	2% (full scale)	2% (full scale)

The various specifications are defined as follows:

Range: Min and max measurement limits

Min detectable sensitivity: Smallest amount of input concentration which can be detected as concentration approaches zero

Rise time 90%: Interval between initial response time and time to 90% response after a step increase in inlet concentration

Fall time 90%: Interval between initial response time and time to 90% response after a step decrease in inlet concentration

Zero drift: Change in instrument output over a stated time period of unadjusted continuous operation when the input concentration is zero

Span drift: Change in instrument output over a stated period of unadjusted continuous operation when input concentration is a stated upscale value

Precision: Degree of agreement between repeated measurements of the same concentration (which shall be the midpoint of the stated range) expressed as the average deviation of the single results from the mean

Operation period: Period of time over which the instrument can be expected to operate unattended within specifications

Noise: Spontaneous deviations from a mean output not caused by input concentration changes

Interference equivalent: Portion of indicated concentration due to the total of the interferences commonly found in ambient air

Operating temperature fluctuation: Ambient temperature fluctuation over which stated specifications will be met

Linearity: Max deviation between actual instrument reading and reading predicted by a straight line drawn between upper and lower calibration points necessary. Proper interpretation and utilization of data collected from well-designed air quality surveillance programs are a matter of utmost significance in the implementation of air quality control.

The methodology outlined in the body of this paper reflects the current approach used by the Tennessee Valley Authority in its program of ambient air monitoring at the agency's coal-fired power plants. It embodies much of what has been learned over a period of more than 20 years in TVA's air quality program.

REFERENCES

1. Montgomery, T. L., Charles L. Massey, Auburn E. Owen, Jr., and William C. Colbaugh. "Surveillance of Air Quality Related to Power Plants and Implementation of the Clean Air Act of 1970," Proceedings of the Winter American Society Mechanical Engineers Meeting, November 29–December 2, 1971, Washington, D.C.
2. Environmental Protection Agency. "Requirements for Preparation, Adoption, and Submittal of Implementation Plans," *Federal Register,* Volume 36, No. 158, August 14, 1971, Washington, D.C.
3. Carpenter, S. B., T. L. Montgomery, Jack M. Leavitt, William C. Colbaugh, and Fred W. Thomas. "Principal Plume Dispersion Models–TVA Power Plants," *J. Air Pollution Control Assoc.,* 21(8), 491 (August 1971).
4. Steven, R. K., and A. E. O'Keefe. "Modern Aspects of Air Pollution Monitoring," *Analytical Chemistry,* 42, 143A (February 1970).
5. Hodge, J. P., Jr. "When Not to Use On-Line Pollution Instruments," *Instrumentation Technology,* 18, 28 (December 1971).

3.
THE FEASIBILITY OF REDUCING AMBIENT AIR CONCENTRATION OF SULFUR OXIDES BY DISPERSION

F. A. Gifford

Air Resources Atmospheric Turbulence and Diffusion Laboratory
National Oceanic and Atmospheric Administration
Oak Ridge, Tennessee

Let me start out by telling you that, when I was asked to prepare this paper, the title and topic were assigned to me. I say this so that the nearest environmental action group doesn't ride me out of town on a rail. It is, of course, true that the atmosphere has great capability to diffuse properties, including sulfur pollution, but not an unlimited capacity or one that works all the time or under adverse conditions. Furthermore, there are some significant questions concerning the rate of removal of airborne sulfur by precipitation, as well as regional and global atmospheric contamination by sulfur aerosol, that are at the moment not entirely resolved. Both of these points should be kept in mind in any assessment of the capability of the atmosphere to reduce tall-stack sulfur emissions to acceptable levels of ambient air concentration. On the other hand, it is equally true that power companies face continued expansion to meet increasing demands for electricity, that for this purpose they have to engineer power systems many years in advance, that large fractions of this power will be supplied by fossil-fuel combustion for several decades, and that we have right now no satisfactory (*i.e.*, reasonably cheap) full-scale systems to remove sulfur from the combustion gases. Both to achieve and maintain presently required air quality levels and to guide the engineering design of future large-scale systems for removal of sulfur from stack gases, the question of atmospheric dilution has to be considered. I will in the following present the essential meteorological facts, emphasizing the importance of understanding certain general facts about tall-stack sulfur pollution.

SOURCES OF ATMOSPHERIC SULFUR POLLUTION

Sulfur is added to the atmosphere by various natural and human activities and removed by various processes. The current global sulfur balance is illustrated in Figure 1, after Robinson and Robbins.[1] From this you see that the total pollutant SO_2 source is a major part of the sulfur balance, and you can compare it with the other sources, sinks, and exchange processes. It is of interest that natural sulfur emissions, mainly as H_2S, are about 1/3 greater

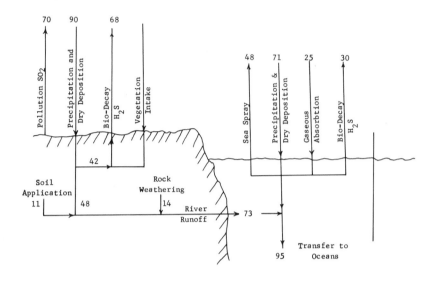

(10^6 tons/year Sulfur)

Figure 1. World Sulfur Balance.

than industrial emissions of SO_2 and H_2S; SO_2 is the only significant industrial pollutant. Note that the oceans' net gain of sulfur is 95×10^6 tons/year, as a result of runoff, rainout and deposition, and gaseous absorption. Three-quarters of this sulfur comes to the oceans from agricultural applications to the soil and weathering of rocks.

Meteorologists in Oak Ridge find it convenient to classify (stationary) pollutant sources into three groups: (1) tall stack sources, (2) process emissions, and (3) area sources. By tall stacks we have in mind stacks in the height range from 200 feet or so up to giants such as the 800-foot Bull Run Steam Plant stack and even taller ones. The main requirement is that the stack should be tall enough, and otherwise well enough designed, so as not to suffer from "downwash" due to the adverse aerodynamic influence of nearby buildings or of inadequate stack draft. By process emissions we mean emissions from lower stacks or vents of industrial processes of all kinds whose height and disposition do not qualify them as tall stacks. And by area sources we have in mind the multitude of individual sources that comprise a city or town, mainly from space heating of all kinds. All three of these can be sources of sulfur pollution, and each requires somewhat different consideration from the meteorological point of view. All three types of sources as a rule occur in conjunction because power demand, industry, and urbanization go together. Mine-mouth power plants are, perhaps, an exception.

Tall-Stack Sources

The behavior of plumes from tall-stack sources, as described by Briggs,[2-4] is dominated by the initial buoyancy of the hot gases emitted. This causes a strong rise of such plumes, initially, accompanied by a bending over due to the (horizontal) ambient wind. The relative motion between the rising plume and the atmosphere causes small-scale turbulence due to velocity shear at the plume's edge and results in a dilution of the initial plume buoyancy by entrainment into the plume of ambient air. The upshot is that tall-stack plumes containing sulfur from fuel combustion rise to great heights, usually more than twice the source height. Stack designers depend on this plume rise to control ground-level air concentrations to acceptable levels. Formulas for calculating plume rise under various meteorological conditions can be found in the papers by Briggs.

We understand plume rise fairly well, but our knowledge of processes that bring the sulfur back down to the ground again is rudimentary. Tall-stack plume behavior is summarized in Figure 2.[2] Plumes are carried back down to the ground briefly under "looping" conditions. Plume "looping" occurs under conditions of strong thermal convective turbulence in the lower atmosphere, similar to the conditions that are accompanied by cumulus clouds. The result at the ground is very brief periods (a few minutes) of high ground level concentrations but a fairly low average concentration value due to the generally good atmospheric dispersion involved. These bursts of high concentration can occur as near to the source as about four times the stack height.

Plumes are also brought to ground level by the so-called fumigation condition. This simply refers to a rapid, general mixing downward of the plume under active mechanical turbulence (high wind) conditions, possibly exacerbated by the presence of a capping temperature inversion aloft. Because this condition can persist for longer periods of time, it produces the highest average ground-level concentrations as a rule.

The normal downward diffusive spreading of plumes can in principle also bring material to the ground. For strongly buoyant plumes, this occurs at great distances from the stack, probably no less than 20 stack heights, which means that considerable dilution will have occurred. The great distance is because downward turbulent mixing can be effective only after the stage of actively buoyant plume rise is over.

Sulfur from plumes can also reach the ground by being absorbed by falling raindrops. The extent and time scale of this phenomenon are quite uncertain. Rain is an efficient scavenger of plume SO_2, but the droplets evidently give up the sulfur readily as they fall below the plume. Thus, the plume sulfur is in effect "smeared out" in the vertical, and the net result of rainfall is to augment vertical SO_2 dispersion. The result for ground level concentrations as well as the extent of so-called acid rain are uncertain. My opinion, based on current, incomplete literature, is that sulfur pollution at the ground attributable to precipitation scavenging of tall stack plumes is not especially

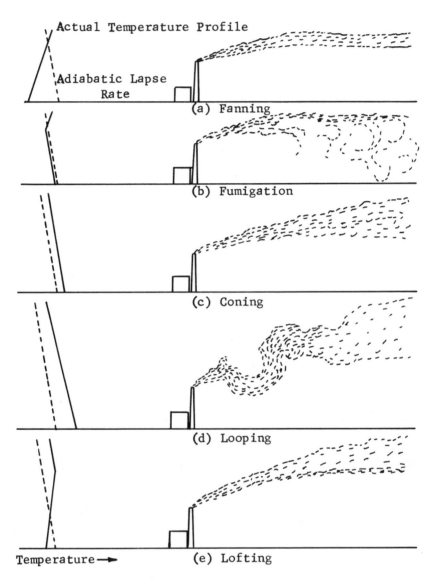

Figure 2. Effect of temperature profile on Plume Rise and Diffusion

intense. That is, the high sulfur concentrations present in the plumes do not result in high ground-level concentrations or intense "acid rain" in precipitation. I would hazard the guess that the plume sulfur reaching the ground from this effect is first fairly well dispersed in the atmosphere by the combined actions of entrainment, turbulent mixing, and the adsorbtion-desorbtion effect. Thus, the plume sulfur should be thought of as tending to contribute to the overall, general sulfur content of rainfall rather than as being a source of highly local "hot spots." I believe that the primary source of higher than average rain and soil acidity buildups, such as has been reported in Sweden, will be found to be area and process emission sources and not tall stacks. I should stress, however, that direct, conclusive evidence for (or against) this opinion does not yet exist.

Process Emission Sources

The problem of process emissions is illustrated by Figure 3, taken from the ASME Guide.[5] Stacks can emit material either above the aerodynamic building wake, within the wake, or, for very short stacks, within the low-pressure cavity in the lee of the building. In the last two cases, ground-level concentrations can be high at all distances near the source. The problem involved in trying to calculate pollution from process emissions is to balance properly the complex plume buoyancy and momentum effects, aerodynamic building wake effects, and the influence of atmospheric turbulence which, near the ground, is highly variable. Guidance can be found in the ASME booklet.

Area Sources

The multitude of small, low-level pollution sources, including residential and industrial space-heating chimneys and other such sources, are most conveniently dealt with as total emissions per unit area. The main difficulty here is to establish the area source strength, and systematic procedures for doing this have been established by HEW. Simplified procedures for calculating area source concentrations can be found in the papers by Gifford and Hanna[12] and Hanna.[13]

THE RELATIVE CONTRIBUTION TO AMBIENT AIR QUALITY OF SULFUR POLLUTION FROM THE VARIOUS SOURCE TYPES

SO_2 from tall stacks emerges high above the ground, then rises still higher because of plume buoyancy, and is only intermittently carried to the ground by air turbulence or precipitation processes. For this reason, tall stacks tend to contribute relatively small amounts of SO_2 to the ambient air concentration at ground level. This was clearly brought out in the recent study by Ross et al.[6] Figure 4 from their paper shows that, while total SO_2 emissions from all sources in UK increased from 1958 to 1968, the annual average concentration decreased. This was due to the controls imposed on the sulfur content of

(a)

(b)

(c)

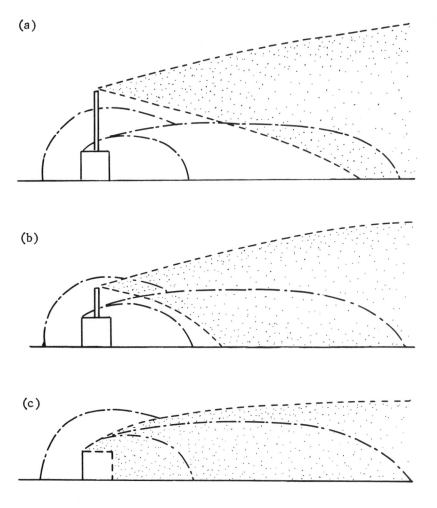

Figure 3. Aerodynamic Effect on Plume Dispersion. (These figures show the behavior of stack plumes emitted on various heights and positions relative to a cubical building.)

fuels for the low level (lower solid curve) process emissions and area sources, mainly domestic heating. A related result was recently reported by Golden and Mongan,[7] who calculated air pollution for Chicago from emissions data. They found that the chief contribution to ambient air SO_2 concentrations is the air source component.

At a smaller, local scale, the Rockwood-Harriman air pollution study recently completed[8] by my colleagues at ATDL produced an equivalent result, as Table 2 shows. The conclusion from all such studies, as argued very

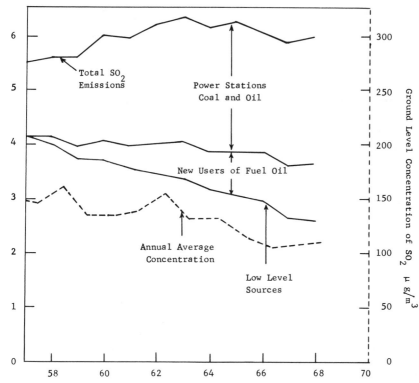

Figure 4. Annual Emissions and Average Annual Concentrations in Atmosphere of U.K., 1957-1968.

Table 2. Suspended Particle Concentration in the Rockwood-Harriman Area

Source	Annual Avg. Amt.
Upwind cities: Nashville, Knoxville, Chattanooga, etc.	- 40 μ/m^3
Kingston Steam Plant	- 5-10 μ/m^3
Roane Electric: in Rockwood	- 50 μ/m^3
in Harriman	- 25 μ/m^3
Space heating, both towns	- 50 μ/m^3

eloquently by Ross *et al.*, is that tall stacks are in fact an excellent and probably an indispensible way to reduce ground level pollution by the emitted material to tolerable levels.

LARGER-SCALE AND GLOBAL CLIMATE EFFECTS

As indicated earlier, sulfur pollution from tall stacks is one of the major contributors to the atmospheric sulfur balance. Some atmospheric SO_2 is quickly removed by precipitation scavenging and other processes, but some becomes oxidized to sulfate aerosol. Such fine particles change the heat balance of the earth[9] by reflection and absorption of both solar and terrestrial radiation. Even such small particles are fairly efficiently removed by precipitation scavenging and other removal processes. Examination of solar radiation records in North America and Europe has led Budyko[10] to speculate that the observed 1940 to 1960 solar radiation decrease of about 4% is due to aerosols from anthropogenic pollution. Mitchell,[11] on the other hand, believes that the manmade aerosol fraction is much less than that from volcanic eruptions. The situation is not particularly clear and should be studied very carefully. The recent MIT summer study[9] concludes that problems in climate modification due to the incerase in aerosol population will tend, because of the somewhat uncertain but relatively short atmospheric residence times of aerosols, to be a regional problem long before they become a global one and, for the same reason, the trend can be reversed quickly, within a few months, if pollution control measures are employed.

SUMMARY AND CONCLUSIONS

1. Tall stacks contribute a sizeable fraction of the total sulfur in the air. With other anthropogenic sources, they contribute an amount comparable with that from natural sources of sulfur. Most of this is SO_2.

2. Sulfur pollution sources are of three types: tall stacks, process emissions, and area sources.

3. Tall-stack sources of sulfur are buoyant and this produces large additional plume rises, essentially doubling the effective release height.

4. Sulfur in tall-stack plumes is brought down to the ground intermittently by atmospheric turbulence and also by precipitation scavenging.

5. The contribution of tall-stack plumes to SO_2 ambient air quality levels averages appreciably less, by a factor of 5 to 10 or more, than that from lower-level process emissions and area sources (where these are present).

6. The possibility of "acid rain" from tall-stack plumes exists. But the (so far rather meager) observational evidence suggests that the contribution will not be especially intense, compared to that from process emission and area sources of sulfur. That is, tall stacks probably should in this respect be thought of as contributing a pro rata share to the general atmospheric sulfur content of a region.

7. Part of tall stack SO_2 emissions are converted to sulfate aerosols. The precise extent and importance of this effect are not known, but, as with the rainfall acidity, any adverse meteorological effects appear to be regional to continental in scale, *i.e.*, neither local nor global.

Acknowledgment
This chapter was originally developed for the U.S. Department of Commerce and is published herein with the permission of the U.S.D.C.

REFERENCES

1. Robinson, E., and R. C. Robbins. "Gaseous Atmospheric Pollutants from Urban and Natural Sources," in S. F. Singer, ed., *Global Effects of Environmental Pollution*. (Berlin: Springer-Verlag, 1970), pp. 50–63.
2. Briggs, G.A. *Plume Rise* (Washington, D.C.: USAEC, DTI, 1969), vi + 81 pp.
3. Briggs, G.A. "Optimum Formulas for Buoyant Plume Rise," Phil. Trans. Roy. Soc. Lond. A. *265*, 197–203 (1969).
4. Briggs, Gary A. "Some Recent Analyses of Plume Rise Observations." Paper presented at the 1970 International Air Pollution Conference of the International Union of Air Pollution Prevention Associations, ATDL contribution number 38, 1970.
5. Smith, Maynard, ed. *Recommended Guide for the Prediction of the Dispersion of Airborne Effluents*. ix + 85 pp., 1968.
6. Ross, F. F., A. J. Clarke, and D. H. Lucas. "Tall Stacks—How Effective Are They?" Proc. of the Second Int. Clean Air Congress of the Int. Union of Air Poll. Prevention Assoc., Dec. 6–11, 1970. (Washington, D.C.: in press.)
7. Golden, J., and T. R. Mongan. "Sulfur Dioxide Emissions from Power Plants: Their Effect on Air Quality," Science *171*, 381–383 (1971).
8. Hanna, Steven R., and S. D. Swisher. "Air Pollution Meteorology of the Rockwood-Harriman, Tennessee Valley." ATDL contribution No. 40, 1970.
9. Massachusetts Institute of Technology. *Man's Impact on the Global Environment* (Cambridge, Mass.: M.I.T. Press, 1970), xxii + 319 pp.
10. Budyko, M. I. "The Effect of Solar Radiation Variations on the Climate of the Earth," Tellus *21* (1969).
11. Mitchell, J. M. "A Preliminary Evaluation of Atmospheric Pollution as a Cause of the Global Temperature Fluctuation of the Past Century," in S. F. Singer, ed. *Global Effects of Environmental Pollution* (Berlin: Springer-Verlag, 1970), pp. 139–155.

12. Gifford, F. A., and Steven R. Hanna. "Urban Air Pollution Modelling." Paper presented at 1970 International Air Pollution Conference of the International Union of Air Pollution Prevention Associations, ATDL contribution No. 37, 1970.

13. Hanna, Steven R. "Simple Methods of Calculating Dispersion from Urban Area Sources." Paper presented April 6, 1971, at the Conference on Air Pollution Meteorology, sponsored by the American Meteorological Society in cooperation with the Air Pollution Control Association, at Raleigh, North Carolina, ATDL contribution No. 46, 1971.

4.
DESCRIPTION OF ATDL COMPUTER MODEL FOR DISPERSION FROM MULTIPLE SOURCES

Steven R. Hanna

Air Resources Atmospheric Turbulence and Diffusion Laboratory
National Oceanic and Atmospheric Administration
Oak Ridge, Tennessee

In order to estimate surface concentrations of pollutants in regions containing many sources, it is convenient to place the emissions into two categories: area sources (*e.g.,* home heating or small industries) and point sources (*e.g.,* power plants or large industries). In previous reports,[1,2] simple methods of calculating concentrations due to area sources were derived and verified. We have devised a straightforward computer model for accounting for point sources as well as area sources. This model has been applied to many regions, including Rockwood-Harriman, Tennessee,[3] and Knoxville, Tennessee.[4] It is the purpose of this paper to outline the ATDL computer program in detail so that others may use it.

CALCULATION OF SURFACE CONCENTRATIONS DUE TO AREA SOURCES

Most source emission inventories distribute the area sources in an urban region into a square grid pattern, with grid distances of about 5 km, where it is assumed that the area source strength in any grid square is uniform across that square. Gifford's[5] "reciprocal plume" concept is employed in order to estimate the surface concentration $\chi(\mu g/m^3)$ due to area sources upwind of the receptor point:

$$\chi = \int_0^D \sqrt{\left|\frac{2}{\pi}\right|} \, \frac{Q_A}{U\sigma_z} \, dx \qquad (1)$$

where $Q_A(\mu g/m^2 sec)$ is source strength, $D(m)$ is the distance to the edge of the urban area, $U(m/sec)$ is the wind speed, and $\sigma_z(m)$ is the vertical dispersion parameter. Smith[6] suggests the following empirical form for σ_z:

$$\sigma_z = ax^b, \quad (x \text{ in meters}) \qquad (2)$$

23

Sunny day	a = .40	b = .91
Cloudy day	a = .15	b = .75 (also applies to yearly averages)
Night	a = .06	b = .71

If the receptor is at the center of grid block "0," with grid distance Δx, and the wind blows in only one direction, then equation 1 can be written as the summation over grid squares upwind of the receptor square:

$$\chi = \sqrt{\frac{2}{\pi}} \; \frac{(\Delta x/2)^{1-b}}{Ua(1-b)} \left[Q_{A0} + \sum_{i=1}^{4} Q_{Ai} \left[(2i+1)^{1-b} - (2i-1)^{1-b} \right] \right] \quad (3)$$

The source strengths $Q_{A0}, Q_{A1}, Q_{A2}, \ldots$, apply to the grid square in which the receptor is located, the grid square upwind of the receptor square, and so on, respectively. The integration is arbitrarily terminated after four grid squares. If the frequency with which the wind blows from the 16 major directions is known, then equation 3 becomes the double summation:

$$\chi = \sqrt{\frac{2}{\pi}} \; \frac{(\Delta x/2)^{1-b}}{Ua(1-b)} \left[Q_A(0,0) + \sum_{i=\pm 1}^{\pm 4} \sum_{j=\pm 1}^{\pm 4} Q_A(i,j) \; f(i,j) \right.$$

$$\left. \left[(2r+1)^{1-b} - (2r-1)^{1-b} \right] \right] \quad (4)$$

where r is the number of grid blocks that square (i,j) is away from the central receptor square. For example, r equals one for the "ring" of eight squares adjacent to the central receptor square. When the 16 point wind direction frequency distribution is input to the program, the parameters $f(i,j)$ and

$(2r+1)^{1-b} - (2r-1)^{1-b}$ are calculated within the program, using techniques

explained by Gifford and Hanna.[1]

In the program, therefore, each element $C(i,j)$ of a 9 by 9 matrix of coefficients is multiplied by the corresponding element $Q_A(i,j)$ of the source matrix to obtain the concentration χ in the central square. Thus, it is necessary to expand the source matrix by 4 squares on all sides, entering small, background source strengths in these squares, so that the 9 by 9 coefficient matrix will "fit" over the source grid along the edges of the given urban area.

These techniques are straightforward and do not necessarily require a digital computer. We have often made them using a slide rule or desk calculator. However, the point source calculations, described in the next section, are sufficiently complicated that a digital computer should be used.

CALCULATION OF SURFACE CONCENTRATIONS DUE TO POINT SOURCES

The basic Gaussian plume formula described by, for example, Slade[7] is used to estimate surface concentrations due to point sources. For a 16 point wind direction frequency distribution, the surface concentration χ of a pollutant emitted at strength Q_p ($\mu g/sec$) is given by the formula:

$$\chi = \sqrt{\frac{2}{\pi}} \; \frac{fQ_p}{\sigma_z r U \frac{2\pi}{16}} \; e^{-\frac{H^2}{2\sigma_z^2}} \tag{5}$$

where H(m) is the effective source height, r(m) is the distance of the receptor point from the source, and f is the frequency with which the wind blows towards the sector of interest. (This formula is equivalent to Slade's[7] equation 3.144.) The effective source height H is the sum of the stack height h_s and the plume rise h_p, which is calculated using the formula proposed by Briggs:[8]

$$h_p = 2.9 \left\{ w_0 R_0^2 \left[T_{po} - T_{eo} \right] \; / \; \left[U \left(\frac{\partial T_e}{\partial z} + .01° \; C/m \right) \right] \right\}^{1/3} \tag{6}$$

where g, T, w_0, and R are the acceleration of gravity, absolute temperature, initial plume vertical speed, and stack radius, respectively. The subscripts p and e refer to plume and environment variables, respectively.

Equation 5 can be used in order to estimate the average concentration in the grid square in which the point source is located. The source is assumed to be located in the center of the square, and the average concentration in the circle with radius $\Delta x/2$ is assumed to equal the average in the square with side Δx. Also since there is only one wind direction sector for the source square, the fraction $2\pi/16$ in equation 5 becomes 2π, and the frequency f equals one.

$$\bar{\chi}_{source \; square} = \frac{1}{\pi \left(\frac{\Delta x}{2} \right)^2} \int_0^{2\pi} d\theta \int_0^{\Delta x/2} \sqrt{\frac{2}{\pi}} \; \frac{Q_p}{ar^b U 2\pi} \; e^{-\frac{H^2}{2a^2 r^{2b}}} \; dr \tag{7}$$

This equation is integrated numerically by the computer, given the input parameters Δx, Q_p, a, b, H, and U. It is found that the average concentration due to a ground level (H = 0) point source located in the middle of the grid square is equal to the average concentration due to a uniform area source in that grid square, provided that Q_p equals Q_A $(\Delta x)^2$; *i.e.*, the total emissions in the square are equal. The average concentration $\overline{\chi}$ in grid squares other than the source grid square is assumed to equal the concentration in the center of the square calculated using equation 5. Gifford's[5] reciprocal plume concept is then used to estimate the total surface concentration in a grid square due to contributions from the given distribution of point sources by superimposing a 9 by 9 matrix over the source grid, and multiplying term by term. This is done in the program by classes of effective stack height, H.

PROGRAM DESCRIPTION

Input

The computer program, written in FORTRAN IV for use on an IBM 360/65 computer, performs the above calculations using the following input parameters. Maximum permissible dimensions and units are given.

Card 1;   (2F10.2,4I5)
DX: grid distance (m)
BX: rural or background source strength ($\mu g/m^2$ sec)
NR: number of rows, including an extra 4 rows on the top & bottom
NC: number of columns, including an extra 4 columns on either side
Nϕ: number of corrections to area source strengths (if no corrections, put Nϕ = 1. If no area sources, put Nϕ = 0.)
NH: number of effective source height classes
Card 2;   (16I5); NN: number of stability classes
Card 3;   (8F10.2); RX(10): parameters a in $\sigma_z = ax^b$
Card 4;   (8F10.2); PX(10): parameters b in $\sigma_z = ax^b$
Card 5;   (8F10.2); U: wind speed (m/sec)
Card 6;   (8F10.2); F(16): wind direction frequency distribution beginning with NNE and going clockwise
Card 7;   (20A4); ITIT1: description of area sources
Card 8;   (20A4); ITIT2: description of point sources
Card 9;   (20A4); ITIT3: description of concentration patterns
Card 10;   (16I5); ID(30): row number of area source correction. If no correction, ID(1) = 1
Card 11;   (16I5); JD(30): column number of area source correction. If no correction, JD(1) = 1
Card 12;   (8F10.2); ST(30): area source corrections ($\mu g/m^2$ sec). If no correction, ST(1) = .0
Card 13;   (8F10.2); S(30,30): area source strengths ($\mu g/m^2$ sec)

Card 14;   (8F10.2); H(20): effective source heights (m)
Card 15;   (16I5); NP(20): number of point sources in each source height
           class
Card 16; ⎫   ⎧(16I5);IC(10,30): row number of point source
Card 17; ⎬   ⎨(16I5);JC(10,30): column number of point source
Card 18; ⎭   ⎩(8F10.2); QEC(10,30): point source strength (μg/sec)
           If no point sources, put NH = 1, H(1) = 1, NP(1) = 1, IC(1,1) =
           1, JC(1,1) = 1, QEC(1,1) = 0.0

Output
Maximum permissible dimensions are given.
1. (8F10.2); F(16): wind direction frequency distribution, beginning with
   NNE and going clockwise
2. (8F10.2): FU(9,9): 9 by 9 matrix of direction frequency distribution,
   divided by wind speed
3. NR (number of rows), NC (number of columns), Nϕ (number of area
   source corrections), NH (number of effective source height classes)
4. DX (grid distance [m]), BX (rural source strength [μg/m^2 sec]), U (wind
   speed [m/sec])
5. R, B: current values in $\sigma_z = Rx^B$, applying to the following numbers 6-13
6. C(9,9): matrix of coefficients for area sources
7. A(30,30): concentrations (μg/m^3) due to area sources
8. H(20): current effective source height, applying to numbers 9-11
9. IC(10,30), JC(10,30), QEC(10,30): current location and magnitude of
   point sources
10. SS(9,9): matrix of coefficients for current point sources
11. AA(30,30): concentrations (μg/m^3) due to current point sources
12. AC(30,30): concentrations (μg/m^3) due to all sources
13. RAT(30,30): ratio of concentration due to area sources to concentra-
    tion due to all sources

EXAMPLE OF OUTPUT OF PROGRAM

Consider the town of Moonshine, Tennessee (population 603), a square
town which has been divided into a square grid system consisting of four 5
km by 5 km grids. Each grid square has an area source emission of 1
μg/m^2 sec. A point source, with effective source height, H = 30m, and source
strength $10^7 \mu$g/sec, is in the center of the southwest block. The wind blows
with equal frequency from all directions; i.e., F(1) = F(2) ... = F(16) = 0.67.
Assume that the wind speed U equals 5 m/sec. Consider only the stability
class applicable to average yearly conditions: NN = 1, RX(1) = .15,
PX(1) = .75.

Other input parameters are then: $DX = 5000.$, $BX = .0$, $NR = 10$, $NC = 10$, $N\phi = 1$, $NH = 1$, $ID(1) = 1$, $JD(1) = 1$, $ST(1) = .0$, $S(5,5) = S(5,6) = S(6,5) = S(6,6) = 1.$, $NP = 1$, $IC(1,1) = 6$, $JC(1,1) = 5$, $QEC(1,1) = 10^7$.

The predicted surface concentrations due to area and point sources are listed in Table 3. These numbers can be used by anyone to check whether or not they are using the program correctly.

Table 3: Source Data and Predicted Surface Concentrations at Moonshine, Tennessee

Grid Square	Area Source $\mu g/sec\ m^2$	Point Source $10^7\ \mu g/sec$	Surface Concentration Due to Area Sources $\mu g/m^3$	Surface Concentration Due to Area Sources $\mu g/m^3$
NW	1.0	0.0	34.64	0.58
NE	1.0	0.0	34.64	0.33
SW	1.0	1.0	34.64	3.70
SE	1.0	0.0	34.64	0.58

APPENDIX: PROGRAM LISTING

```
      DIMENSION S(30,30),FU(30,30),C(30,30),A(30,30),QE(30,30),D(30,30),
     1SS(30,30),AA(30,30),AC(30,30),IC(10,30),JC(10,30),QEC(10,30), ID(3
     20),JD(30),ST(100),RX(10),PX(10),ITIT1(20),ITIT2(20),ITIT3(20),H(20
     3),NP(20),AAB(30,30),RAT(30,30),F(16)
100 FORMAT(2F10.2,4I5)
105 FORMAT(8F10.2)
110 FORMAT(20A4)
163 FORMAT(16I5)
      READ(5,100)  DX,BX,NR,NC,NO,NH
      DO 300 I=1,NR
      DO 300 J=1,NC
300 S(I,J)=BX
      READ(5,163)  NN
      READ(5,105)  (RX(I),I=1,NN)
      READ(5,105)  (PX(I),I=1,NN)
      READ(5,105)  U
      READ(5,105)  (F(I),I=1,16)
      READ(5,110)  ITIT1
      READ(5,110)  ITIT2
      READ(5,110)  ITIT3
      READ(5,163)  (ID(I),I=1,NO)
      READ(5,163)  (JD(J),J=1,NO)
      READ(5,105)  (ST(J),J=1,NO)
      JI=NC-4
      IJ=NR-4
      READ(5,105)  ((S(I,J),J=5,JI),I=5,IJ)
      READ(5,105)  (H(I),I=1,NH)
      READ(5,163)  (NP(I),I=1,NH)
      DO 860 I=1,NH
      NNP=NP(I)
      READ(5,163)  (IC(I,J),J=1,NNP)
      READ(5,163)  (JC(I,J),J=1,NNP)
      READ(5,105)  (QEC(I,J),J=1,NNP)
```

```
 860 CONTINUE
     WRITE(6,794) (F(I),I=1,16)
 794 FORMAT(2X,9HWIND ROSE/(8F10.4))
     DO 789 I=1,9
     DO 789 J=1,9
 789 FU(I,J)=.0
     E=1./U
     FU(5,5)=E
     DO 790 I=1,4
     FU(5-I,5+I)=F(2)*E
     FU(5+I,5+I)=F(6)*E
     FU(5+I,5-I)=F(10)*E
     FU(5-I,5-I)=F(14)*E
     FU(5-I,5)=F(16)*E
     FU(5,5+I)=F(4)*E
     FU(5+I,5)=F(8)*E
 790 FU(5,5-I)=F(12)*E
     FU(4,5)=FU(4,5)+E*(F(1)+F(15))
     FU(5,6)=FU(5,6)+E*(F(3)+F(5))
     FU(6,5)=FU(6,5)+E*(F(7)+F(9))
     FU(5,4)=FU(5,4)+E*(F(11)+F(13))
     DO 791 I=1,2
     FU(1+I,6)=F(1)*E
     FU(4,6+I)=F(3)*E
     FU(6,6+I)=F(5)*E
     FU(6+I,6)=F(7)*E
     FU(6+I,4)=F(9)*E
     FU(6,1+I)=F(11)*E
     FU(4,1+I)=F(13)*E
 791 FU(1+I,4)=F(15)*E
     FU(1,7)=F(1)*E
     FU(3,9)=F(3)*E
     FU(7,9)=F(5)*E
     FU(9,7)=F(7)*E
     FU(9,3)=F(9)*E
     FU(7,1)=F(11)*E
     FU(3,1)=F(13)*E
     FU(1,3)=F(15)*E
     WRITE(6,9) ((FU(I,J),J=1,9),I=1,9)
   9 FORMAT(2X,37HINPUT WIND SPEED AND FREQUENCY MATRIX/(9F10.5))
     WRITE(6,470) NR,NC,NO,NH
 470 FORMAT(2X,9HNO. ROWS=,I5,3X,12HNO. COLUMNS=,I5,3X,3HNO=,I5,3X,21HN
    10 EFFECT SOURCE HTS=,I5)
     WRITE(6,10) DX,BX,U
  10 FORMAT(2X,9HDX IN M =,F6.0,3X,23HRURAL SOURCE STRENGTHS=,F5.1,3X,1
    11HWIND SPEED=,F10.3,5HM/SEC)
     DO 600 I=1,NR
     DO 600 J=1,NC
 600 QE(I,J)=.0
     DO 167 K=1,NO
     I=ID(K)
     J=JD(K)
 167 S(I,J)=ST(K)
     WRITE(6,200) ITIT1,((S(I,J),J=1,NC),I=1,NR)
 200 FORMAT(2X,20A4/(10F10.2))
     DO 162 II=1,NN
     R=RX(II)
     B=BX(II)
     WRITE(6,119) R,B
 119 FORMAT(2X,43HNEW SET OF POWER LAW PARAMETERS FOR SIGMA Z/3X,2HR=,F
    15.3,3X,2HB=,F5.3)
     IF(NO) 160,160,150
 150 CONTINUE
     DO 1 I=1,9
     DO 1 J=1,9
   1 C(I,J)=.0
     BB=((DX/2.)**(1.-B))/(R*(1.-B))
     CC=9.**(1.-B)-7.**(1.-B)
     DD=.80*BB*CC
     DO 2 J=1,9,2
     C(1,J)=DD*FU(1,J)
   2 C(9,J)=DD*FU(9,J)
     DO 3 I=3,7,2
     C(I,1)=DD*FU(I,1)
   3 C(I,9)=DD*FU(I,9)
     CC=7.**(1.-B)-5.**(1.-B)
     DD=.80*BB*CC
```

```
      C(8,8)=DD*FU(8,8)
      C(2,2)=DD*FU(2,2)
      C(2,8)=DD*FU(2,8)
      C(8,2)=DD*FU(8,2)
      DO 4 J=4,6
      C(2,J)=DD*FU(2,J)
   4  C(8,J)=DD*FU(8,J)
      DO 5 I=4,6
      C(I,2)=DD*FU(I,2)
   5  C(I,8)=DD*FU(I,8)
      CC=5.**(1.-B)-3.**(1.-B)
      DD=.80*BB*CC
      DO 6 J=3,7
      C(3,J)=DD*FU(3,J)
   6  C(7,J)=DD*FU(7,J)
      DO 7 I=4,6
      C(I,3)=DD*FU(I,3)
   7  C(I,7)=DD*FU(I,7)
      CC=3.**(1.-B)-1.
      DD=.80*BB*CC
      DO 8 J=4,6
      C(4,J)=DD*FU(4,J)
   8  C(6,J)=DD*FU(6,J)
      C(5,4)=ID*FU(5,4)
      C(5,6)=DD*FU(5,6)
      C(5,5)=.80*BB*FU(5,5)
      WRITE(6,11) ((C(I,J),J=1,9),I=1,9)
  11  FORMAT(2X,39HMATRIX OF COEFFICIENTS FOR AREA SOURCES/(9F10.5))
      DO 500 I=5,IJ
      DO 500 J=5,JI
      A(I,J)=.0
      DO 500 K=1,9
      DO 500 L=1,9
 500  A(I,J)=C(K,L)*S(I-5+K,J-5+L)+A(I,J)
      WRITE(6,95)
  95  FORMAT(2X,34HCONCENTRATIONS FROM GROUND SOURCES)
      WRITE(6,110) ITIT3
      WRITE(6,700) ((A(I,J),J=5,JI),I=5,IJ)
 700  FORMAT(8F10.4)
 160  IF(QEC(1,1)) 162,162,161
 161  CONTINUE
      WRITE(6,202)
 202  FORMAT(2X,16HELEVATED SOURCES)
      WRITE(6,110) ITIT2
      DO 899 I=5,IJ
      DO 899 J=5,JI
 899  AAE(I,J)=.0
      DO 880 NI=1,NH
      DO 874 KL=1,NR
      DO 874 KM=1,NC
 874  QE(KL,KM)=.0
      WRITE(6,871) H(NI)
 871  FORMAT(2X,26HEFFECTIVE SOURCE HEIGHT = ,F10.3)
      WRITE(6,872)
 872  FORMAT(10X,6HROW NO.3X,9HCOLUMN NO.4X,25HEMISSION(MICROGM PER SEC)
     1)
      NNP=NP(NI)
      WRITE(6,873) ((IC(NI,K),JC(NI,K),QEC(NI,K)),K=1,NNP)
 873  FORMAT(2I12,F20.5)
      DO 164 K=1,NNP
      I=IC(NI,K)
      J=JC(NI,K)
 164  QE(I,J)=QEC(NI,K)
      DO 81 L=1,9
      DO 81 K=1,9
  81  D(I,K)=0.0
      XY=H(NI)**2./2.
      P=EXP(-XY/(R**2.*((DX/2.)**(2.*B))))
      PA=R*((DX/2.)**(B+1.))
      XINT=.0
      DDR=.0
      XX2=DX/2.
      DO 948 IJJ=1,100
      IF(IJJ-10) 952,952,953
 952  DR=100.
      GO TO 954
```

```
953 DR=1000.
954 DDDR=DDR+DR/2.
    DDR=DDR+DR
    XYZ=XY/(R**2.*(DDDR**(2.*B)))
    IF(XYZ-20.) 960,960,961
960 XINT=XINT+DR*EXP(-XYZ)/(DDDR**B)
961 CONTINUE
    IF(XX2-DDR) 955,955,948
955 IJJ=100
948 CONTINUE
    D(5,5)=XINT/(8.*R*XX2**2.)
    DO 71 L=1,4
    E=2*L
    EF=E**(2.*B)
    XF=1.4*E
    FE=XF**(2.*B)
    Z=P**(1./EF)/(PA*(E**(B+1.)))
    IF(L-1) 950,950,949
949 D(5,5+L)=Z
    D(5,5-L)=Z
    D(5+L,5)=Z
    D(5-L,5)=Z
    GO TO 951
950 D(5,6)=Z/3.
    D(5,4)=Z/3.
    D(6,5)=Z/3.
    D(4,5)=Z/3.
951 Z=P**(1./FE)/(PA*(XF**(B+1.)))
    D(5+L,5+L)=Z
    D(5-L,5+L)=Z
    D(5+L,5-L)=Z
 71 D(5-L,5-L)=Z
    DO 72 L=1,2
    QQ=2*L+2
    QG=1.1*QQ
    G=QG**(2.*B)
    Z=P**(1./G)/(PA*(QG**(B+1.)))
    D(4-L,6)=Z
    D(4-L,4)=Z
    D(6+L,4)=Z
    D(6+L,6)=Z
    D(4,4-L)=Z
    D(6,4-L)=Z
    D(4,6+L)=Z
 72 D(6,6+L)=Z
    Z=P**(.125**(2.*B))/(PA*(8.**(B+1.)))
    D(1,3)=Z
    D(1,7)=Z
    D(3,1)=Z
    D(3,9)=Z
    D(7,1)=Z
    D(7,9)=Z
    D(9,3)=Z
    D(9,7)=Z
    DO 91 K=1,9
    DO 91 L=1,9
 91 SS(K,L)=2.04*D(K,L)*FU(K,L)
    WRITE(6,92) ((SS(I,J),J=1,9),I=1,9)
 92 FORMAT(2X,33HCOEFFICIENTS FOR ELEVATED SOURCES/(9F12.11))
    DO 510 I=5,IJ
    DO 510 J=5,JI
    AA(I,J)=.0
    DO 410 K=1,9
    DO 410 L=1,9
    ADD=SS(K,L)*QE(I-5+K,J-5+L)
    AA(I,J)=AA(I,J)+ADD
410 AAB(I,J)=AAB(I,J)+ADD
510 CONTINUE
    WRITE(6,96) H(NI)
 96 FORMAT(2X,50HCONCENTRATIONS FROM ELEVATED SOURCES WITH HEIGHT =,F1
   10.3)
    WRITE(6,110) ITIT3
    WRITE(6,700) ((AA(I,J),J=5,JI),I=5,IJ)
880 CONTINUE
    IF(NO) 166,162,166
```

```
166 CONTINUE
    IF(NH) 169,162,169
169 CONTINUE
    DO 520 I=5,IJ
    DO 520 J=5,JI
    AC(I,J)=A(I,J)+AAB(I,J)
520 RAT(I,J)=A(I,J)/AC(I,J)
    WRITE(6,97)
 97 FORMAT(2X,31HCONCENTRATIONS FROM ALL SOURCES)
    WRITE(6,110) ITIT3
    WRITE(6,700) ((AC(I,J),J=5,JI),I=5,IJ)
    WRITE(6,890)
890 FORMAT(2X,65HRATIO OF CONCENTRATION DUE TO AREA SOURCES TO TOTAL C
   1CNCENTRATION)
    WRITE(6,700) ((RAT(I,J),J=5,JI),I=5,IJ)
162 CONTINUE
    STOP
    END
```

REFERENCES

1. Gifford, F. A., Jr., and S. R. Hanna, 1970: Urban air pollution modelling, presented at 1970 Meeting of the Int. Union of Air Poll. Prev. Assoc., Washington, Dec. 11, 1970, 17 pp + 7 figs.
2. Hanna, S. R., 1971: A simple method of calculating dispersion from urban area sources, *J. Air Poll. Control Assoc., 21,* 774-777.
3. Hanna, S. R. and S. D. Swisher, 1970: Air Pollution Meteorology of the Rockwood-Harriman, Tennessee, Industrial Corridor. Atm. Turb. and Diff. Lab. Rept. 40, Box E, Oak Ridge, Tennessee, 21 pp.
4. Hanna, S. R., 1972: An Air Quality Model for Knox County, Tennessee. Atm. Turb. and Diff. Lab., Rept. 55, Box E, Oak Ridge, Tennessee, 34 pp.
5. Gifford, F. A., 1959: Computation of pollution from several sources. *Int. J. of Air Poll., 2,* 109-110.
6. Smith, M. E. (Editor), 1968: Recommended guide for the prediction of the dispersion of airborne effluents, ASME., ix and 85 pp.
7. Slade, D. (Editor), 1968: *Meteorology and Atomic Energy, 1968,* USAEC Rept. No. TID-24190, x and 445 pp.
8. Briggs, G. A., 1969: *Plume Rise,* USAEC Rpt. No. TID-25075, vi + 82 pp.

Acknowledgment

This research was performed under an agreement between the National Oceanic and Atmospheric Administration and the U.S. Atomic Energy Commission.

5.
EVOLUTION OF AIR POLLUTION EMISSION STANDARDS FOR STATIONARY SOURCES

Kenneth E. Noll

Associate Professor, Civil Engineering
The University of Tennessee

A typical set of air pollution control regulations used by control agencies today is both complex and voluminous. The comprehensive nature of these regulations has been necessitated by the growing complexity of air pollution control in our modern society. This paper provides a historical perspective regarding current regulations within a generalized framework in which the numerous regulations can be organized and understood.

An emission standard is a limit on the amount of a pollutant which can be emitted from a source and is intended to bring the concentration of the pollutant in the ambient air within acceptable ambient air quality standards. Ambient air quality standards are maximum desired levels of pollution in the outdoor air, established by governmental agencies as legal regulations. Air quality standards are in turn based on air quality criteria which are scientific statements of the relationship between dosage and adverse effects of pollutants.

Figure 5 shows the interrelationship between the various standards and indicates the sequence of events, with alternates, which can be used to establish emission limit standards on sources of pollution. The figure shows that air quality standards are based on air quality criteria. The data collection stage (air monitoring, emission inventory, and meteorological data) is required so that some type of relationship can be established between presently existing air quality and the desired air quality levels as set by air quality standards. This relationship may be established in a number of ways (different models), but the results allow for some rational scheme for determining the amount of air pollution reduction required (*i.e.,* roll-back, proportional reduction, or diffusion model). This in turn allows specific emission standard levels to be established on sources.

The above procedure is certainly not the only rational way in which specific emission limits can be established, although it seems to be the most scientifically and technically sound. There are at least three other procedures which can be used. The first of these alternate methods requires the application of the best available technology which will result in minimum discharge with no regard for present air quality. The second provides for the direct

33

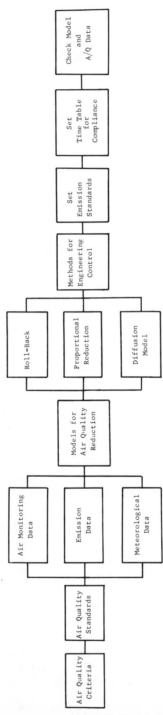

Figure 5. Control Steps in Air Pollution Engineering

elimination of effects such as odors, irritation, or visibility reduction. The third provides no specific emission limits but specifies that ground-level ambient air quality standards near single sources of pollution not be exceeded.

CATEGORIES FOR EMISSION STANDARDS

The evolution of air pollution emission standards for stationary sources has been going on for over 70 years. The number and type of emission standards have necessarily increased as (1) air pollution control has become more complex, (2) many new substances have been considered pollutants, (3) many new types of industries have evolved.

Also, as our understanding of air pollution and its effects has increased, there has been a definite tendency to make air pollution control standards more specific and more restrictive. Emission standards have also been in a constant state of change due to at least three specific items: (1) lower level and more numerous air quality standards, (2) larger urban areas and industrial operations, and (3) development of better technology for control.

There also has been a trend toward the development of emission standards which are more restrictive for larger size operations. This tendency has been due in part to the much larger plant size of today with the resulting potential for the release of large amounts of material at one location and also to the fact that there is a definite economy of scale involved in the control of large sources of pollution as compared to smaller ones. Thus, it is usually cheaper to treat gas streams on a dollar per standard cubic foot of stack gas as the amount of gas which requires treatment increases.[1]

Even though a great number of diverse emission control standards have been developed and selected ones have been published as examples of good regulations,[2] they can be categorized into seven general areas as follows:

1. Visible emission standards.
2. Particulate matter concentration standards.
3. Exhaust gas concentration standards.
4. Prohibition of certain types of emissions.
5. Performance standards.
6. Regulation of fuel.
7. Zoning restrictions.

The following sections will provide specific examples of each of these type emission control regulations and information on the application and significance of each standard will also be provided.

Visible Emission Standards

The most widespread type of control regulation is based on the Ringelmann chart and is aimed at controlling the emission of visible material from a source. Basically, this is a method of judging the shade of gray of a given smoke plume. The Ringelmann chart consists of five dotted or cross-hatched sections corresponding to different percentages of black. When the chart is displayed between an observer and a gray smoke plume, the shades of gray on the various sections can be compared with the shade of the smoke. Thus, Ringelmann Number 1 is equivalent to 20% black, Ringelmann Number 2 to 40% black, Ringelmann Number 3 to 60% black, and Ringelmann Number 4 to 80% black.

The Ringelmann chart was developed in the 1890s by Maximilian Ringelmann[3] and was introduced into the United States around 1900. It began to appear frequently in control regulations after this date. Present regulations are based on U.S. Bureau of Mines publications.[4]

There has been a great deal of controversy over the accuracy of this method, which has extended over the entire history of this regulation as seen by trial records of 1900, publications of the 1930s,[5] and discussions of today. Most of the conflict is caused by the subjective nature of this standard, requiring trained observers for legal enforcement. One of the main objections to the use of the Ringelmann scale—that there is no general correlation between plume opacity and the quantity of pollutants released—has been somewhat overcome by recent studies by Pilat.[6]

The popularity of the visible emission regulation stems from the fact that it offers a cheap and relatively simple means of enforcement. A trained observer can take many observations of a plume, and it is much less costly than stack sampling and analysis. Access to the plant premise is not required, and an observer can maintain surveillance of a wide area.

In 1947, the Los Angeles Air Pollution Control Agency first applied the concept of equivalent vision obscuration to nonblack plumes.[7] Today, many ordinances have an "equivalent opacity" clause which allows a white plume to be compared to an equivalent obscuring power on the Ringelmann chart.

The majority of smoke emission laws prohibit the emission of a Ringelmann number or equivalent opacity darker than Number 2 (40%), but the trend today is toward the Number 1 (20%) level[8]. At least one regulation provides for no visible emissions from new sources.

Regulations based on plume opacity are often difficult to comply with because they require the removal of small particles in the range of 0.1 to 1.0 microns. These particles are responsible for light scattering but usually contribute little to the total weight of material in a plume. Thus, when a weight based standard is attained, many small particles, which are more difficult to remove, may still be emitted and cause a visible plume.

Particulate Matter Concentration Standards

There are two types of regulations for the control of particulate matter which have traditionally been used. One type is based on the weight concentration of particles (mass emission) in the stack and the other relates the weight of emitted particles to the total weight of material processed (process weight).

Mass Emission Standard

Mass emission standards usually limit particulate matter in a stack to some upper limit such as 0.2 grains/STD. cu. ft. (7000 grains/lb.). This value may be different for different type operations such as incinerators, wood-waste burners, blast furnaces, open hearth furnaces, cupolas, and other metallurgical operations.

Mass emission regulations are designed to apply to combustion sources, where it is possible to correct the concentration to some standard combustion condition such as a certain percent excess air, CO_2, or O_2 in the exhaust gas.

This type of standard may also be written in terms of the total rate of heat input. A sliding scale is usually provided which requires more control for larger units (Figure 6). This type of sliding scale also allows a different standard for different sources. For example, waste-wood burners and small incinerators could have a heat input below 10 million BTU per hour and a maximum emission of 0.6 lb. per million BTU while large furnaces would be allowed only 0.2 lb. per million BTU.

The mass emission standard originated with the American Society of Mechanical Engineers[9] and was developed during the 1940s to apply to coal-fired boilers. The ASME model law specified that for combustion operations dust loading should not exceed 0.85 lbs. of dust per 1,000 lbs. of stack gas adjusted to 50% excess air or 12% carbon dioxide.

In newer regulations,[7] there has been a general tendency to eliminate the confusion between combustion and other type operations by applying the regulation to all sources. In actual practice, however, present mass emission standards are less restrictive on noncombustion sources and compliances with the process weight rule will generally provide compliance with the mass emission limit.

Process Weight

Rules based on process weight limit the dust emission on the basis of the pounds of material processed and thus preclude the use of dilution as a means of meeting a particulate emission limit. In this approach, permissible emission rates are related to material processed by a sliding scale as shown in Figure 7.

This standard usually governs the emissions from dust-producing operations such as cement plants, rock-crushing operations, and lime kilns. For this regulation, a typical definition of process weight is "the total weight of all material introduced into any specific process which may cause any discharge into the atmosphere." Solid fuels are usually considered as part of the process but liquid and gaseous fuels are not.

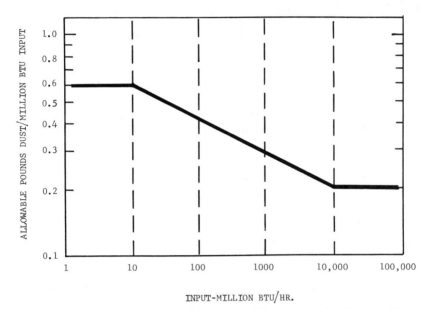

INPUT-MILLION BTU/HR.

Figure 6. Allowable Emissions of Particulate Matter from Fuel-burning Equipment

The process weight rule was developed for Los Angeles County in the late 1940s.[10] The upper limit on the Los Angeles regulation (Figure 7) of 40 lbs/hr, regardless of the size of the plant, has led many persons to think that the process weight rule is not a sliding scale but provides a limit of 40 lbs/hr. Figure 3 shows that in many regulations the emission limit is allowed to continue upward, but at a lesser rate, for operations with a process rate larger than 60,000 lbs/hr.

The original regulation allowed for approximately 80% control for small sources, 90% control for medium size processes, and as much as 99.9% for very large plants. There appears to be some tendency today for process weight rates to be tailored to specific industries instead of setting one rate for all of them. For example, it would be possible to have individual process weight rates for steel mills, cement plants, and metal operations. This trend is due to the fact that the degree of difficulty of control is not the same for each type of source.

It should be stressed at this point that the three particulate matter emission regulations (visibility reduction, mass emission, and process weight) provide control agencies with three complimentary regulations. All of these can be applied to any specific source of pollution. However, in actual practice one of the regulations will be more restrictive as to the required degree of control for a specific source and this regulation will govern the abatement control program.

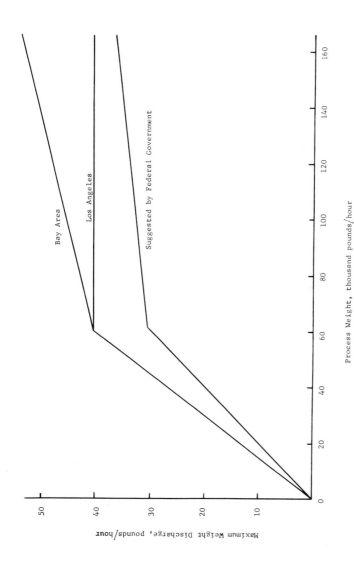

Figure 7. Typical Process Weight — Maximum Discharge Relationship

Exhaust Gas Concentration Standards

Control of gaseous pollutants has been much less widespread than the control of particulate matter. However, considerable attention has been given to the control of two specific gases, sulfur oxides and fluoride compounds. For example, a typical SO_2 regulation requires that no person shall cause, suffer, allow, or permit the emission from any source gases containing more than 2000 parts per million of SO_2. The trend today is toward stricter regulations on sulfur compounds, with limits near 300 to 500 parts per million.

Fluoride limits are usually required to prevent accumulation in vegetation. Hydrocarbon control can take the form of vapor pressure standards for the control of evaporation or vapor recovery system specifications. Odor control can be obtained by limiting the concentration of emission or by requiring special controls such as the burning of gases from rendering plants.

The trend for the future is to control the emission of more gases. The oxides of nitrogen have been singled out for specific regulations in the last few years, and the control of hazardous material will generate new gas emission concentration standards.

Prohibition of Certain Emissions

Some types of activities or operations may be completely banned due to the severity of the air pollution which they produce. Thus, open burning restrictions prevent inefficient disposal of solid waste. The prohibition usually encompasses all open burning from those at dumps and land-clearing projects to the home burning of leaves. Single chamber incinerators may also be banned because of the large amount of pollution which they produce.

A general nuisance regulation is usually included in most comprehensive regulations such as "a person shall not discharge such quantities of air contaminants which cause injury, detriment, nuisance, or annoyance to any considerable number of persons." Most of the early air pollution lawsuits were filed under the nuisance regulation. In fact, problems with these cases provided the stimulus required for more specific regulations.

Performance Standards

Performance standards usually require that processes operate so as not to exceed air quality standards in the surrounding environment without specification of a maximum emission rate. This allows for the use of buffer zones and tall stacks for dilution.

The standard developed by the Bay Area Air Pollution Control District in the late 1950s is a good example of a performance standard.[11] It provides for maximum permissible SO_2 ground-level concentrations at the boundary of the plant. The regulation requires continuous monitoring of ground level concentrations and specifically states that the requirements do not apply to points on any property controlled by the person responsible for the emissions. Another type of regulation requires that the owner or operator of a source emitting sulfur oxides demonstrate by ambient air monitoring that the source does not contribute to ground-level concentrations of sulfur oxides in excess of 80% of the secondary ambient air quality standard.

There are numerous examples of performance type regulations which allow greater emission rates from tall stacks because of the dilution of pollution with clean air before it comes into contact with ground based receptors. Some of these allow calculations to be made on classical dispersion equations and some provide tables and figures where the calculations have already been completed. The following is an example of the formula type regulation which allows a choice between the process weight regulation and the diffusion equation.[12]

For those owners or operators electing to have emissions regulated by diffusion equations, the maximum allowable particulate emission shall be determined by procedures defined below in subsection A, B, or C.

A. Stack Gas Exit Temperature Less Than 100° F

$$A = 3.02 \times 10^{-4} V_s h_s^2 \left[\frac{(d_s)}{(h_s)} \right]^{0.71}$$

B. Stack Gas Exit Temperature of 125° F or Greater
  1. Stacks less than 500 feet
$$Q = 0.2 h_s \left[Q_T \times 0.02 \times (T_s - 60) \right]^{0.25}$$

  2. Stacks 500 feet and greater

$$Q = 0.3 h_s \left[Q_T \times 0.02 \times (T_s - 60) \right]^{0.25}$$

C. Stack Gas Exit Temperatures from 100° F to 124° F

Calculate allowable emission as in A and either B1 or B2, depending upon stack height (using T_s of 125^OF) and make linear interpolation based upon actual stack gas exit temperature.

The terms of the preceding equations shall have the following meaning and units:

d_s = inside diameter or equivalent diameter of stack tip in feet
h_s = stack height in feet (vertical distance above grade directly below tip of stack)
Q = maximum allowable emission rate in pounds per hour
Q_T = volume rate of stack gas flow in cubic feet per second calculated to 60^OF

T_s = temperature of stack gases at stack tip in $^{\circ}F$
V_s = velocity of stack gases at stack tip in feet per second

The trend in the United States is definitely away from allowing higher emission rates for higher stacks. This trend seems to be based on global scale weather changes such as large acid rain belts.[13] Even though the diffusion formula and sliding scale emission rates with stack height are going out of favor in the United States, many foreign control agencies rely heavily on this method for reducing ground-level concentrations.

Another type of performance emission standard which has been used extensively in air pollution control is often referred to as design standards. The requirements that combustion equipment, process operation, and air pollution control equipment meet minimum standards has led to the submission of plans and specifications to control agencies for comparison with design standards concerned with such things as furnace volume and type, chimney height, and fuel type.

Another performance standard is designed to control odors from rendering plants and requires the incineration of off gases at a minimum temperature of $1200^{\circ}F$ for at least 0.3 sec.

Regulation of Fuel

Fuel standards may limit the kind of fuel that may be burned in equipment of specified design, provide for fuel dealer licensing, require fuels to be used from approved sources, or control the content of the fuel.

Many older regulations (starting as early as 1900) contained restrictions on the use of high volatile bituminous coal in certain types of boilers. Some regulations also limited or prohibited the use of solid fuels or required the use of coke.

In the late 1950s, Los Angeles County required that fuel with a sulfur content in excess of 0.5% by weight could be burned only between May 1 and September 30 each year. Today the sulfur content of fuel is limited in many areas. Thus, a limit of 0.5% fuel sulfur content is becoming commonplace in large metropolitan areas.

Another example of fuel specifications is the regulation of the olefin content of gasoline by Los Angeles County to reduce reactive hydrocarbons in the atmosphere.

Zoning Restrictions

Most air pollution rules and regulations do not specifically contain zoning restrictions. However, zoning restrictions may be employed to prevent certain types of new industry from locating in areas where present air quality standards are exceeded or where sensitive or unusual receptors may be located nearby.

Most of the zoning emission standards were developed during the 1950s. These standards either classify industries and then provide location zoning based on the classification system, or allow emissions based on a pounds per hour per acre rate which is designed to incorporate the buffer zone concept.

Air pollution and urban planning agencies now appear to be coming to the realization that urban form—the spatial and temporal arrangement of urban elements—is responsible for present broad-scale urban air pollution problems. This type of problem is much more difficult to control than the single source situation. Future regulations will require concepts concerned with urban form to be truly effective in controlling air pollution problems.

REFERENCES

1. Edmisten, N. G., and F. L. Bunyard. "A Systematic Procedure for Determining the Cost of Controlling Particulate Emissions from Industrial Sources," *Journal Air Pollution Control Association 20* (7), 446 (July 1970).
2. A Compilation of Selected Air Pollution Emission Control Regulations and Ordinances, U.S.P.H.S. Publication No. 999-AP-43 (1968).
3. Ringelmann, M. "Methode d'Estimation des Fumes Produites par Les Foyers Industriels,"*La Revue Technique 268* (June 1898).
4. U.S. Bureau of Mines Information Circular, 7718 (1955).
5. Marks, L. S. "Inadequacy of the Ringelmann Chart," *Mechanical Engineering 214*, 681 (September 1937).
6. Pilat, M. J., and D. S. Ensor. "Plume Opacity and Particulate Mass Concentration," *Atm. Envr. 4*, 163 (1970).
7. Rules and Regulations, Los Angeles County Air Pollution Control District (March 3, 1967).
8. Stern, A. C. "Air Pollution Standards," in A. C. Stern, ed., *Air Pollution Control*, Vol. III (New York: Academic Press, 1968), p. 601.
9. Recommended Guide for the Control of Dust Emission, ASME Standard, American Society of Mechanical Engineers (1966).
10. Yocom, J. E. "Air Pollution Regulations—Their Growing Impact on Engineering Decisions," *Chemical Engineering* (July 1962).
11. Bay Area Air Pollution Control Board Regulation 2, San Francisco, California (January 1962).
12. Tennessee Air Pollution Control Regulations, Tennessee Department of Public Health, Nashville, Tennessee, 1972.
13. LiKens, G. E., F. H. Bormann, and H. M. Johnson. "Acid Rain," *Envr. 14* (2), 33 (March 1972).

6.
LAND USE CONTROL: AIR POLLUTION CRITERIA FOR PLANT SITE SELECTION

Aelred J. Gray

Associate Professor, Graduate School of Planning
The University of Tennessee

Other contributors to this volume deal with the specific air quality standards for industry and the hardware necessary to meet these standards. This paper will be limited to a consideration of some of the requirements for public control and guidance of industrial location; specifically, it will deal with the air pollution criteria useful to relate industrial plant location to city and regional plans and to some of the methods available for public control over such locations.

Air pollution standards as a part of the plant site selection process are important considerations in any approach to the overall problem of planning for a living environment. It is assumed at the start that in the South in general and Tennessee in particular the transition from an agricultural to an urban industrial economy is now essentially complete. The following data suggest the speed with which this transformation has taken place. In 1940, 44% of the people in Tennessee lived and worked on farms. By 1970, this percentage had dropped to 10.

While there have been many adjustments in population distribution and employment as a result of these changes, the urbanization pattern in the Southeast and in Tennessee is still in its formative stages. What would seem to be emerging are systems of cities focused on the region's metropolitan centers. In Tennessee, for example, Nashville is ringed by a system of cities located 25 to 50 miles out. There are similar groupings of cities in the Great Valley of East Tennessee around Chattanooga, Knoxville, and the tri-cities of Kingsport, Johnson City, and Bristol.

The process of city building has always challenged man's imagination. The industrial revolution produced the smoke-filled, dirty towns which pushed man to search for approaches to city building. The objective was to produce a sound and healthful environment for the growing number of people living in cities and towns. It was during this period that an obscure clerk in the British civil service by the name of Ebenezer Howard wrote the book entitled *Tomorrow*. The ideas outlined in this book became incorporated into the "new town movement." Howard proposed that city size be controlled and that new urban growth be directed to entirely new cities. The cities would be interconnected by rail and highway systems and would be separated by open space to provide air, light, and a green countryside as a complement to the dense development in the cities.

In America, the 1893 Chicago World's Fair stimulated interest in cities and city building and the problems associated with even-then sprawling metropolitan areas. The Fair emphasized the city beautiful, with emphasis on recreation, open space, and the separation of residential and industrial areas.

These lessons were forgotten as people by the millions moved off the farms and into American cities. New cities were built, but they were unplanned. More and more people moved to existing cities. Almost overnight the United States became a nation of city dwellers. The city represented power, the new technology, innovation, progress, and new ideas; and, more important, the jobs were there.

Because the cities were built without adequate attention to a living environment, once again—as at the turn of the century—there is a public concern with the air and water pollution and other environmental problems in urban areas. Industrial location is an important part of that concern.

PAST APPROACHES TO PLANT LOCATION

Traditionally, industrial plants have been sited principally on the basis of physical characteristics of the land—level terrain, good drainage, a soil that would provide adequate foundation conditions, and access to water, transportation, and utilities. If these physical characteristics were satisfied, casual attention was given to the city or regional plan where one existed.

On the other hand, the city builders and those responsible for preparing city and regional plans tended to relegate industry to land for which there was little or no use. Occasionally lip service was given to wind directions and to the location of industry in relation to residential growth. Overall, both the industrial locaters and the planners gave less thought to planning for existing and future industrial areas than was given to the average residential subdivision.

CHARACTERISTICS OF INDUSTRIAL PLANTS

In considering how industrial siting might be related more specifically to air pollution criteria, the locational characteristics of industrial plants are important. While each plant has specific requirements, studies have been made of certain general plant characteristics. In 1970, TVA, in cooperation with the industrial development groups in middle Tennessee, compiled data on 254 manufacturing plants in 23 middle Tennessee counties.[1] The study covered factors which influence and contribute to the space requirements of modern factories. For example, over 60% of these plants have less than 100 employees per shift. Slightly more than 80% of these plants had less than 100,000 square feet under roof, and 80% of these buildings covered less than a fifth of the plant site. Of the plants, 75% had less than 25 employees per acre. Thus, we have the picture of major industrial plants as being relatively small in size and number of employees and for the most part located in a kind of parklike environment, utilizing 20% or less of the industrial site itself. Large plants do occur, but these are the exception, and because of their large size they present special conditions.

Industrial Location and Land Use Planning

As a part of the changing community attitude about environmental problems, there are already indications that in some communities industry is greeted with protest signs instead of welcome mats.[2] There has been also a notable change in the concerns for relating industry to city and regional plans. As one recent article stated, "Evidence is mounting that urban air pollution problems are caused not only by the presence of such things as smokestacks and automobiles but by the overall spatial and temporal arrangement of these urban elements—in short, by urban form."[3]

As is well known, once pollutants are added to the air, the microclimate affects their dispersal and transportation. This involves consideration of such items as wind speed and the vertical temperature gradient. Studies now under way suggest that, insofar as city building is concerned, some of these elements may not be beyond human control. For example, the city profile itself—the arrangement, shape, and height of structures—actually will affect wind speed and the air pollutant accumulations over the built-up areas. We are learning some of the effects of topography and its relation to temperature and wind speed through studies which show that buildings located on a crest of a hill are likely to experience fewer problems of air pollution than those concentrated in the valley floor where the cooler air settles and where the air pollutants collect. Other aspects of the city form which affect air pollution occur where buildings form man-made canyons which produce wind eddies of great volume that will pick up and transport dust and other air pollutants over the city.

As noted earlier, one of the earliest strategies suggested by planners for improving urban environment was the location of industry downwind from the residential areas. The theory here was that the pollutants from industrial location would not drift into the residential areas.[3] While such devices as stack-cleaning equipment can now control the gravity fallout of many types of emission, these control devices are costly and only partially effective. Other techniques must be used as complements. One such technique is the use of location itself. Identifying and shunning those locations, such as valley areas, where accumulations can be hazardous is an example. The Donora pollution tragedy was partly caused by improper use of such locations. These sites can be used, but the cost of adjusting local conditions is great.

The health and environmental hazards are not limited to closed valleys. Where plants are upstream from a residential area, there should be concern not only for the topographic conditions that exist but also the wind direction and velocity. In all of these circumstances, the microclimate conditions are affected by the orientation and shape of buildings as well as by stack heights.

Rydell and Schwarz, in their review article relating to air pollution and urban form, described good and bad examples of land use planning drawn from Stalingrad (Russia) and Linz (Austria):

> Stalingrad offers the most clear-cut example of designating the macro form of an urban area to minimize air pollution costs. Taking advantage of a wind that almost always blows from the same direction, the planners of this new city organized the major land uses in strips perpendicular to the wind direction. The prevailing wind passes over the residential, recreational, and park areas and only when it is beyond these zones does it accumulate major pollutants by reaching the highway, railroad, and industrial land uses. This macro form also has the virtue of providing short trips to work.
>
> Linz, Austria, however, provides an example of improper land use planning. Because the prevailing wind comes from the west, Linz was planned with residential areas to the west and industrial sites to the east. The attempt to have the wind blow pollution away as in Stalingrad largely failed, however, because the mild east wind banks the pollution against the western mountains around this valley city. As a result the residential area is often blanketed by smoke. The failure of the Linz plan was due to an incorrect understanding of the relationship between wind and air pollution concentrations. High velocity winds cause rapid vertical and horizontal dispersion of pollution, while low velocity winds slowly carry pollution through nearby areas causing high pollutant concentrations. Wind frequencies must be weighted by the inverse of wind speed before they can be used to locate residential areas out of the path of industrial pollution.[3]

Others suggest more emphasis on the control of population and building densities. Spreading development over larger areas tends to dilute the air pollutants, and thus concentrations of pollutants will tend to decrease. While such an approach might help with some of the immediate air pollution problems, it is also important to keep in mind that continued scatteration of urban development could lead to an increase in other forms of pollution relating to water, soil erosion, and solid waste disposal and, in addition, could add significantly to the cost of providing urban services.

Here again, urban form becomes important because we can encourage the continued large concentrations of people in a relatively few metropolitan areas or we can try to build open space into our city pattern by encouraging systems of small cities. Some of these could be entirely new cities as proposed in recent studies of requirements for a national urban development policy.

ZONING AS ONE TOOL FOR AIR POLLUTION CONTROL

Another approach is based upon the idea of air as a resource and its allocation to specific uses. Building on this concept, air zoning is introduced as a method of dealing with resource allocation. This method would seek to predict the course of air masses carrying pollutants and to allocate air to the users along its path. Certainly, before this kind of control could be utilized, there must be acceptable air quality criteria—acceptable in the sense of the general public—before there can be a justifiable alteration of the spatial location and timing of urban activities.[3]

Utilizing the zoning device as an air pollution control measure is based on the principle which establishes that property-right ownership includes the use of space immediately above the property as well as the land itself. With air rights, one of the rights held in ownership of real property, it seems reasonable that laws regulating land use can also regulate the use of the air space above the land.[4]

Most present-day zoning ordinances do not contain provisions aimed at maintaining a desirable level of air quality. Early zoning theory was built on the principle of concentrating like uses in the same areas. Beginning in 1950 and supported largely by the work of the late Dennis O'Harrow, who for many years was Executive Director of the American Society of Planning Officials, the idea of including performance standards in zoning ordinances began to evolve. This proposal suggested specifying levels of smoke, dust, dirt, and gases permissible within a given district as well as standards for noise, use density, and traffic concentration. Initially, most of these standards applied only to the industrial districts. It was thought that to broaden these standards to include residential areas poses many problems. For example, if the same performance standards prevailing in certain light industrial districts were applied to the single-family district, all window air conditioners and power mowers would not conform to the noise standards, and the emission from many residential areas generally would be illegal.[4]

Recent tendencies have been to categorize the emissions of particulates according to the type of land use for which they originate. The current view seems to be that the same emission densities for all types of land use are impractical and, if applied to industrial location, would have the immediate effect of excluding most industry. It is argued that present-day industrial areas have large amounts of open spaces and are removed from the populated areas and that emissions are generally discharged from elevated stacks. In contrast, the emissions from residential areas are generally discharged from much lower elevations and in more crowded situations. For this reason, it is generally accepted that the requirements for air quality in industrial zones need not be as strict as those in residential areas. Table 4 shows the 1967 allowable particulate emission densities by land use districts in the St. Louis central urbanizing area. Note that the allowable emission densities for industrial areas are about five times greater than for residential areas.[5]

Table 4: Allowable and Existing Particulate Emission Densities by Land Use in St. Louis–East St. Louis Central Urbanized Area

Land Use	Area mi^2	% Total	Existing Emission Density tons/mi^2/yr	Allowable Emission Density tons/mi^2/yr	Additional Reduction Needed, %
Industry	36	9.1	1,890	600	68.2
Residential	153	38.8	170	130	22.3
Commercial	23	5.8	255	175	31.6
Open space	182	46.3	35	100	0.0
Central urbanized area	394	100.0	269	175	39.8

Source: *Interstate Air Pollution Study, Phase II Project Report, VIII A Proposal for an Air Resource Management Program,* U.S. Department of Health, Education, and Welfare, Public Health Service, May 1967, p. 19.

JOINT ACTION BY AIR POLLUTION CONTROL AND CITY AND REGIONAL PLANNING AGENCIES

These new developments suggest the need for an expanded working relationship between the air pollution control agencies and the city and regional planning agencies in order to minimize the air pollution problems and the land use, air use conflicts. Many air resource management plans now include land use planning and zoning as elements for plan implementation.[4] Certainly, the city and regional planning agencies will have to rely on the air pollution agencies for the air quality standards to be included in the zoning ordinances and for the criteria and expertise to enforce them. Kurtzweg described the problem in this way:

In the recent air resource management plans, land-use zoning in conjunction with emission standards has been proposed to achieve a pattern of emission densities that will not create ambient pollutant concentrations exceeding community air quality goals. Essentially, this concept would mean maintaining the intensity of human activity in an area at levels such that the pollutant emissions from these activities, when reduced to the extent technologically and economically feasible, would not over-burden the dispersive capabilities of the atmosphere. Although the performance standards in some zoning codes have been specified in terms of emissions per areal unit of land, there generally has not been any rigorous relationship between these standards and the potential future ambient concentrations. Not infrequently the performance standards have been lifted *en toto* from the zoning code of an area with a completely different type of emission sources and meteorologic and topographic characteristics. The role the air resource management proposals indicate for zoning, while not entirely different from that which it has played before, will require the introduction of new considerations in the development of the comprehensive zoning plan.

For the first time, a close working relationship between the air pollution control agency and the planning agency seems possible. The funds available for local and regional air pollution control agencies have increased rapidly during the last decade. For example, the non-Federal budgeting for air pollution control amounted to $26.8 million as of April 1969 compared to only $9.3 million in 1961.[6] As the activities of these agencies expand, city and regional planners will have more help in determining quantitative standards needed for better land use planning.

CONCLUSION

It seems reasonable to state that we are only now beginning to understand some of the interrelationships between air pollution and land use control. While the attempt at air pollution control through land use may not be the ultimate answer, it can help to relate emissions to our present living environment. Ultimately, every effort must be made to internalize costs to achieve low emissions.

The current rule of thumb for industrial locations seems to be to look for sites in industrial zones where other plants are already well established. This suggests that more attention should probably be given to the establishment of industrial parks where wastes, including those relating to particulate emissions, can be treated through some kind of centralized system much like our existing city sewerage works. These kinds of questions are forcing the industrial site location people, city and regional planners, and air management staffs to talk together and to try to understand and relate their different points of view.

Finally, there seems to be a growing concern about the size of future cities. Funds are being provided and assistance is available to many small communities for housing, water, sewerage, schools, hospitals, and roads. Perhaps by reducing our concentrations of people in individual areas and by providing the facilities whereby industry can be dispersed without economic penalty, we can utilize the great open spaces of this country to help to cut down on the serious pollution problems that now threaten our cities and their related regions.

REFERENCES

1. *Site and Employment Characteristics of Selected Manufacturing Plants in Middle Tennessee*, Division of Navigation Development and Regional Studies, Tennessee Valley Authority, December 1970.
2. "New Headache for Site Seekers: Public Clamor Over Pollution," *Chemical Week*, August 19, 1970.
3. Peter Rydell and Gretchen Schwarz, "Air Pollution and Urban Form: A Review of Current Literature," *Journal of the American Institute of Planners*, March 1968.
4. J. A. Kurtzweg, "Land Use Planning and Air Pollution Control in the Puget Sound Basin," paper presented at the Annual Meeting of Pacific Northwest Section, Air Pollution Control Association, Salem, Oregon, November 8–10, 1967.
5. *Interstate Air Pollution Study, Phase II Project Report, VIII A Proposal for an Air Resource Management Program*, U.S. Department of Health, Education and Welfare, Public Health Service, May 1967.
6. Thomas W. Ellington, "Air Pollution Control: A 'State of the Art' Report," *Water and Sewage Works*, February 1970.

Section II

Power Generation and Air Pollution

7.
POWER GENERATION AND THE ENVIRONMENT

W. B. Harrison

Southern Services, Inc.
Birmingham, Alabama

In the following comments, an attempt is made to present a view of various elements which influence the present and future growth of the electric utility industry. Quite obviously, environmental protection is one of the more important elements. These comments are based on personal opinion in some instances, rather than hard facts, but, where possible, an effort is made to present evidence on which the opinions are based.

The dynamic nature of modern society is such that social, political, moral, technical, and economic forces interact in a complex manner which influences every major enterprise of man. The electric utility industry is one of man's largest enterprises, and its use of resources, its mode of operation, and its interaction with the natural environment make it highly visible to the public at all times. For the same reasons, it must be responsive to priorities which are set by the public in various ways, but it necessarily has technological and economic constraints which influence both the scope and direction of feasible alternatives for change. It is hoped that these comments will resolve some of the uncertainty which appears to be developing with respect to three factors: (a) demands created by regulations which are presumed to reflect the will of the public; (b) well-defined requirements for protection of public health and welfare, based on factual evidence; and (c) feasible alternatives from the point of view of available technology and related costs.

DEMAND PROJECTIONS

Forecasting always has uncertainties which must be recognized at the outset. In the case of electrical demand projections, the forecasting technique which appears to be most used is simple trend extrapolation. In approaching the extrapolation for demand, the principal input information is in two parts: (a) the anticipated population growth and (b) the trend in per capita use of electricity.

Of first importance is the population trend. Looking back at the U.S. population growth over the last twenty or thirty years gives some evidence about what the future growth may be like. However, one must recognize that there are significantly changing attitudes in the United States concerning family size. The demographers, who make their professional mark by studying such questions, point out that even a difference as small as three-tenths of a child in the average family size, or three children in every ten families, would make a difference of as much as 25 million in the total population by the year 2000.[1] The forecasts made only three or four years ago have been revised downward more recently in order to reflect these changing attitudes.[2,3] However, the overall trend is very clear: continued growth is inevitable. No matter how effective the Zero Population Growth (ZPG) Movement may be, it can have very slight influence over the population in the next decade, and, even taking possible effects into account, the total population growth expected in the United States over the next thirty years will probably amount to some 70 or 75 million persons. If this growth were to be steady, it would be necessary to build the equivalent of a new city of about 200,000 inhabitants every month for the next thirty years. Think of all the requirements of a city of 200,000—housing, transportation, recreation, employment, health services, sewage plants, waste disposal systems, shopping centers, and the variety of products and services needed to construct and support these facilities. Consideration of this growing population is basic as one attempts to assess the challenge of the future.

Next, consider the trend in per capita use of electricity. The best estimates available predict that the rate of growth of per capita use will be impressive (Figure 8).[4-6] Note that this growth is expected in the face of a vocal sector of the population which advocates reduction in the use of electricity as a national goal, presumably motivated by concern for the environment.

There are several factors which influence this forecast for growth. An improvement in average income will mean that an increasingly large sector of the population may acquire the comforts and conveniences made possible by electricity. As noted above, population growth will require new jobs and services, and electricity will provide an increasing share of the related energy needs because, for many applications, it will be cleaner and easier to manage at the point of use than any competitive energy sources. The concern for environmental protection will also accelerate this shift from other energy sources to electricity, and pollution abatement efforts of other industries will undoubtedly create additional demand for electricity.

With regard to industrial use of electricity, it is apparent that increases in productivity will be required to offset escalating labor costs if domestic inflation is to be slowed and if U.S. goods are to be competitive in the world markets. It seems that innovative uses of electricity and strenuous efforts to keep cost of electricity within reason offer the best chance of bringing this about.

To get back to subjects more familiar, note the changing life style in the United States which will have an impact on the use of electricity. Consider things such as completely enclosed, air-conditioned, well-lighted shopping malls, improved lighting on streets and parking areas (creating the combined advantages of improvements in esthetic appeal while reducing crime and

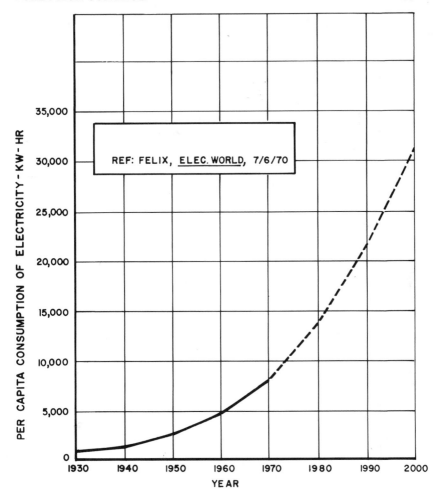

Figure 8. Per Capita Growth Rate in Use of Electricity.

accident rates), air conditioning in homes and offices and the like. Think also of other needs of existing communities which have increasing demands for energy, such as sewage disposal and mass transportation. Since the automobile is a big factor in urban air pollution, the pressures for electric mass transportation systems and an electric commuter car continue to mount. In terms of the energy requirements of society in general and the desire to clean up the environment in particular, it is clear that the per capita use of electricity must increase if a number of national goals are to be met.

Using reasonably conservative figures for population growth and per capita use of electricity, one may confirm the popular forecast that total electrical demand will double during this decade (Figure 9). This forecast also suggests

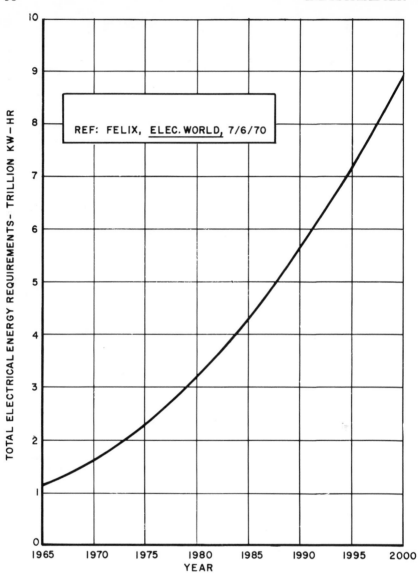

Figure 9. Total Electrical Demand Growth Rate.

that, after 1980, about thirteen years will be required for it to double again in order to reflect the expected decline in the population growth rate. One conclusion from this forecast is that, in the present decade, the electric utility industry faces the most difficult task in its history if it is to achieve the required growth rate in the context of rapidly changing technical, social, economic, and political conditions.

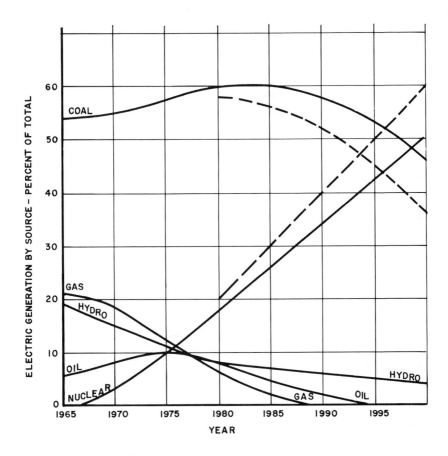

Figure 10. Principal Energy Resources.

ENERGY RESOURCES

Where will the energy come from? The principal energy resources for power generation are water for hydroelectric generation, the fossil fuels—gas, oil, and coal—and nuclear fuels (Figure 10).[7,8] Though some new hydroelectric generation facilities will be built, the possible sites are limited by nature in terms of the terrain and the amounts of water available at any particular place, and therefore, hydroelectric generation will obviously represent a diminishing fraction of the total. Natural gas, for all practical purposes, is disappearing from the utility marketplace. The shortage of natural gas is well documented, and the gas which is available will be used for specialized markets. There are good prospects for the production of synthetic *natural gas* from coal which offer promise to the same specialized markets, like home owners, but these developments are not promising to the utilities because synthetic natural gas is expected to be too expensive for the economical generation of electricity except in special cases. Oil has an uncertain future because reserves are limited and a great portion of the oil used in the United States is from foreign sources which, in turn, are vulnerable and susceptible to international complications of various kinds.[9,10] It is likely that oil will follow somewhat the same fate as natural gas in the utility marketplace, but somewhat displaced in time. The remaining fossil fuel is coal, of which there are abundant reserves, and it seems obvious that coal surely must be one of the principal competitors for meeting future energy demand in the United States.

Nuclear energy is clearly the other competitor, so that, to meet the expected needs, there will be tremendous growth possibilities emerging for both coal-fired and nuclear-powered plants (Figure 11). Estimates depicting this competition for the future make it clear that as gas, oil, and hydroelectric generation diminish, coal and nuclear generation must make up this loss. Further, if nuclear energy cannot achieve the expected growth, coal must make up any deficiencies. This point is made, not to imply any lack of confidence in the technology of nuclear power or its potential importance in the future, but to reflect the uncertainty in achievable rates of growth as a result of changing regulations and adversary actions which seem to be arising throughout the country and which can only have an adverse effect on growth rates. The merits of nuclear energy can withstand the critical appraisal of any reasonable person and emerge as an indispensable, safe, reliable means for electric generation with minimum environmental impact.[11-13] In fact, in considering the needs of future communities, it is clear that nuclear power offers distinct advantages over fossil plants for urban locations because of the relatively simple fuel supply logistics and the freedom from gaseous effluents. Urban locations may also offer new opportunities for utilizing part of the energy, now wasted, for certain community services such as sewage treatment. Regardless of the short-term skirmishes with certain environmentalists and other nuclear power adversaries, the need for and the advantages of nuclear power will surely prevail in the long term. It should also be recognized that nuclear power is caught up in the current of technological change which will surely lead to the successful development of the breeder reactor for electric generation. This reactor, as the name implies, actually produces more nuclear fuel than it consumes. The electric utility industry is now involved in a joint venture with the federal government to expedite development of a breeder reactor power system. Success of this venture will hasten the full utilization of nuclear energy and extend nuclear fuel resources by many orders of magnitude. As shown by the forecasts, however, the growth in energy demand simply cannot be met by nuclear power alone, even under the most favorable circumstances. Accordingly, ways must be developed to use coal in the context of the expected future growth.

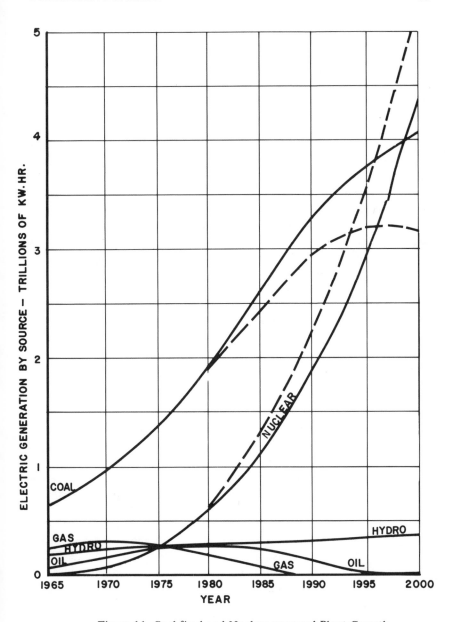

Figure 11. Coal-fired and Nuclear-powered Plant Growth.

ENVIRONMENTAL CONSTRAINTS

The Environmental Protection Agency has officially declared power plants to be stationary pollution sources, and it has further identified three pollutants associated with power generation which must be abated in connection with national air quality goals: particulate matter, sulfur dioxide, and oxides of nitrogen. Other constraints are imposed on the addition of heat to public waters.

The Environmental Protection Agency has adopted two general kinds of regulations concerning air pollution. Ambient air quality standards pertain to the upper concentration limits of various pollutants in the air at ground level, regardless of source. "Primary" standards are related to "health" and "secondary" standards are related to "welfare," including esthetics. In addition to and regardless of the strategy for complying with ambient air quality standards, any new plant for which major procurement agreements were reached on or after August 17, 1971, must also comply with emission limits. These ambient air quality standards and emission limits are summarized on Table 5 as they pertain to power plants.

In general, water temperature criteria limit the maximum temperature to 90° F and the maximum change in the average temperature of a stream to 5° F. In estuaries, the region of temperature change is not well defined, but the limit is 1.5° F.

In order to analyze the environmental effects of power generation, it is convenient to consider a power plant as an isolated sphere in space and observe all streams which enter and leave this sphere. Fuel enters; combustion or reaction products, electricity, and heat leave. There are other minor streams, of course, but the streams noted are the ones of concern in this discussion. When coal is the fuel, the combustion products include all three of the pollutants mentioned above: particulate matter, sulfur dioxide, and oxides of nitrogen. Particulate matter arises from incombustible materials in the coal and it appears as very fine particles, the consistency of face powder, called fly ash. Sulfur dioxide arises because coal contains sulfur in varying amounts and it is converted into sulfur dioxide in the combustion process. Oxides of nitrogen arise from nitrogen in the coal and in the air which is required to support combustion of the coal. Heat is a by-product of all steam-electric generation, conforming to the basic laws of thermodynamics, and it must somehow be assimilated by the environment. Industry research programs are dedicated to developing an understanding of the generation and ultimate fate of these effluents from power stations, and possible technology or strategy for abatement. The following sections deal with effluent considerations in greater detail.

HEAT REJECTION

There are natural and technical constraints which must be recognized in the mode of operation of a power plant facility, and any efforts to

Table 5. EPA Air Quality Emissions Standards

Pollutant	Basis	National Ambient Air Quality Standards		Emissions Standards
		Primary	Secondary	(Max. 2-hr. Average)
Particulate	Annual geo. mean	75 ug/m^3	60 ug/m^3	0.1 lb/MBTU
	Max. 24-hr. conc. (once/yr.)	260 ug/m^3	150 ug/m^3	
SO_x (measured as SO_2)	Annual arith. mean	80 ug/m^3 0.03 ppm	60 ug/m^3 0.02 ppm	Coal, 1.2 lb/MBTU Oil, 0.8 lb/MBTU
	Max. 24-hr. conc. not to be exceeded more than once/yr.	365 ug/m^3 0.14 ppm	260 ug/m^3 0.1 ppm	
	Max. 3-hr. conc. not to be exceeded more than once/yr.		1300 ug/m^3 0.5 ppm	
NO_x (expressed as NO_2)	Annual arith. mean	100 ug/m^3 0.05 ppm	100 ug/m^3 0.05 ppm	Coal, 0.70 lb/MBTU (except lignite) Oil, 0.30 lb/MBTU Gas, 0.20 lb/MBTU

Definitions: ug/m^3, micrograms per cubic meter
ppm, parts per million
MBTU, million British thermal units of heat input

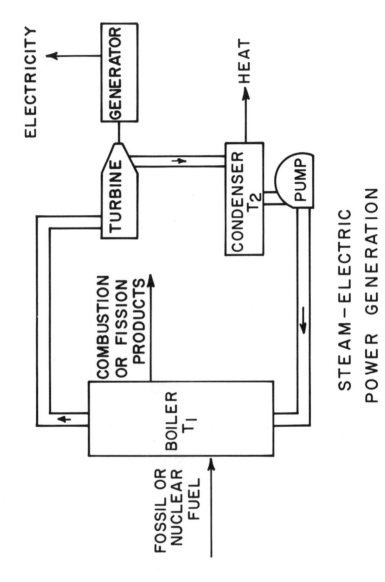

Figure 12. Basic Flow Diagram for Steam-electric Power Generation.

manipulate one of these will have consequences on the others. In a broader context, power generation must be addressed as a part of an overall response to needs of society, including resource conservation, and, in so doing, the public interest can be optimized. Such a concept requires considerable flexibility in the establishment of regulations, for it becomes necessary to appraise every power plant with respect to a variety of consequences from different alternative actions in order to choose the best operating mode to accomplish the optimum benefit to the public.

Figure 12 represents the process which is basic to steam-electric power generation. The key elements shown in this diagram are typical whether the plant derives its energy from fossil fuels (such as gas, oil, or coal) or nuclear fuels. In either case, the boiler is utilized to convert energy from burning fossil fuels or fissioning nuclear fuels in order to produce heat which, in turn, is absorbed by water, converting it from a liquid state to steam at high pressure and temperature. This high-pressure, high-temperature steam is then expanded through a turbine, and, in this process, it provides motive power which drives the electrical generator, producing the electricity which is then transmitted to points of use. As the steam passes through the turbine, it is greatly reduced in temperature and pressure, and it next goes to a condenser from which additional heat must be removed in order to convert the steam from the vapor state back again to a liquid state. The water is then pumped as a liquid back to the boiler and the cycle begins again. In a thermodynamic sense, there are basic requirements for this process. Energy must come into this system as combustion energy or fission energy. Heat must be applied to the working substance, the water, in order to achieve useful power for purposes of driving the turbine and electricity generator. Finally, heat must be rejected from the condenser in order to complete the cycle and restore the water to its original state. For purposes of illustration, temperature T_1 is defined as the characteristic temperature of the boiler and temperature T_2 as the characteristic temperature of the condenser. As a consequence of the laws of thermodynamics, the overall efficiency of this cycle is proportional to $(T_1 - T_2)/T_1$ or $1 - (T_2/T_1)$, where efficiency is defined as the amount of work produced or the amount of electricity generated, compared to the amount of energy entering the cycle. Another way of analyzing this is to note that the difference between the amount of work extracted and the amount of energy entering the cycle must show up as heat rejection. Thus, one may also say that the amount of heat rejection is proportional to T_2/T_1. Note that as T_1 increases, efficiency increases, when T_2 is a constant value. This explains why efficiency of fossil plants can be higher than for most nuclear plants in today's technology since temperatures within a fossil plant boiler can be higher than corresponding temperatures within most nuclear plants. The increased efficiency in the fossil plant can then be interpreted as a reduced or decreased requirement for heat rejection when compared with most nuclear plants with the same electrical output. When T_1 is a constant value, efficiency is increased as T_2 is decreased. Consequently, as T_2 decreases, the amount of heat rejection decreases. Thus, in a power station, the temperature T_1 is usually set as high as experience and available materials will permit so that the magnitude of T_1 is fixed. It follows that one strategy for improving the efficiency in a power plant is to reject heat at the lowest possible temperature, T_2. This explains the practice of the past to locate power plants on natural bodies of water so as to have access to the lowest temperature possible in a particular location.

For many situations, it is both possible and desirable to draw cool water from a natural water body, pass it through a condenser where it picks up energy from condensing steam, and return the warmed water to the water body from which it came. This procedure is referred to as "once-through cooling." It is the cheapest way to provide heat rejection, and it is the most efficient way in the sense that it provides the lowest temperature for heat rejection.[14] In order for this to be a feasible choice, one must consider the amount of water available at a site and the relative amount of water required for condenser cooling as well as the ultimate maximum temperature which this heat rejection will produce in the receiving water. Clearly, there are temperatures above which one cannot go without affecting some change in the aquatic life in the receiving water within the region influenced by the discharge.

Perhaps the most ideal case for once-through cooling utilizes a deep lake. Unfortunately, the possible sites for deep lakes are very limited in number, but, in this application, cold water at the bottom of a deep lake can be used to provide low temperature condenser water, yielding a maximum plant efficiency, and the warm water returned to the lake can have a beneficial effect on the aquatic life in the lake. A shallow lake offers a different set of problems. The land area requirements may range from one to three acres per megawatt of power, and the commitment of land resources must be appraised in analyzing the merit of such a heat rejection strategy.

When once-through cooling is judged to be undesirable, for any reason, the alternatives must be examined. One of these is a cooling tower, and there are two types of cooling towers in use today, both of which are dependent on the evaporation of water to achieve a cooling effect. The difference between these two types arises in the nature of the forces which require movement of the outside air through the cooling tower. In the so-called mechanical draft towers, air is forced through the tower by large fans, providing intimate contact with sprays of warm water. Evaporation of a portion of the warm water gives a cooling effect to the remaining water which returns to the condenser stream. The amount of water evaporated in this process must be replaced from some other source and this is referred to as "make-up water." The principal effluent from the cooling tower is a mixture of warm air and water vapor. In the case of a natural draft cooling tower, the flow of air is induced through buoyant forces arising because the mixture of warm air and water vapor is less dense than the surrounding air. The principal of operation is similar to a mechanical draft tower since the cooling effect is created by evaporating water. The disadvantages of cooling towers of both kinds are (a) the possibility of a nuisance from ground fog which might develop from the water vapor plume; (b) the evaporative losses which are appreciable, and, in areas in which there is a shortage of fresh water supplies, this becomes an important factor; (c) where this cooling water is salty or brackish, finely divided droplets can be carried into the air stream and deposit salt residue downwind from the cooling tower; and (d) they are expensive. These disadvantages must be analyzed with respect to any advantages that are gained by use of cooling towers.

In some cases, the use of sprays can provide an alternative for heat rejection which combines some features of cooling towers and cooling lakes. The cooling effect is provided by evaporation, as in a cooling tower, but there is no confining structure to channel the air flow. Spraying the warm water provides a favorable condition for evaporation and, in so doing, greatly reduces the land area required for the cooling lake. Thus, there are savings in capital cost compared with either cooling towers or cooling lakes, and the operating costs will be less than for cooling towers. There are some situations in which the value of the water resource is so great that the evaporative losses could not be tolerated either in the conventional cooling towers, spray ponds, or large lakes. In such a case, the only alternative remaining would be the dry cooling towers in which heat is exchanged directly from the condenser water to the atmospheric air. This procedure obviously would eliminate the evaporative losses, but would add greatly to projected cost of electricity because of increased capital cost and decreased plant efficiency. At this time, dry cooling is more theoretical than practical, for the required technology is not available.[15] Several references are available which examine these alternatives in detail. Woodson[16-18] has made careful comparisons of once-through cooling, using oceans, rivers, and lakes, and mechanical and natural draft cooling towers, wet and dry, for both fossil and nuclear generation. He used an 800 mw reference plant and assumed typical weather conditions so as to actually reduce the comparisons to effects on plant efficiency and total cost. The exact numbers would not necessarily apply to any particular site, but they are indicative of real effects and real costs. The condenser costs associated with cooling towers are higher than for the once-through system because cooling tower cycles are less efficient and more heat must be rejected. Considerable auxiliary power is required for cooling towers, and this requires an increase in generating capacity in order to produce the net reference generation. This, in turn, requires an additional capital investment in the plant. There are operating cost increases which must also be analyzed to measure the full financial impact of such a choice.

Lee[19] and Kolflat[20] have presented persuasive arguments concerning the merit of once-through cooling to open water and the special virtues of surface discharges at maximum release temperatures in order to accelerate the rate of heat dissipation from the water to the atmosphere and to stratify the zone of influence so as to keep the subsurface regions undisturbed. Using an example of a fixed total heat rejection, Lee[19] illustrated the differences in cooling surface requirements in open water for temperature rises in condenser water of 5, 12, 18, and 27° F, respectively. Flow rates were adjusted to correspond to these temperature rises, and the surface area requirements to return the temperature to near equilibrium conditions were estimated. To return the temperature to within 1° F of the equilibrium temperature, twice as much surface was required for the 5° rise as for the 18° rise. Thus, arbitrary limitations of temperature rise can have an adverse effect on water surface area requirements as well as cost, for pumping costs obviously go up as allowable temperature rise goes down.

In connection with evaporative losses of cooling water, Hauser and Oleson[21] made comparisons of various heat rejection alternatives for a variety of variables, including wet bulb temperature, relative humidity, cloud cover, wind speed, and cooling range. Cooling lakes are shown to be high in evaporative losses because the entire loss from the surface area is charged against the heat rejection task; for a natural river or lake, only the abnormal evaporation is noted. In consideration of water as a limited resource, the losses accompanying heat rejection alternatives are of great importance.

Though heat rejection practices now in use in most installations are adequately safeguarding the aquatic life, it is clear that power plants of the future will not be privileged to adopt practices which have been quite satisfactory in the past.[17,22]

Another aspect of heat rejection is that there are special situations in which the heat can be used to advantage. One example is the use of the warm water discharge for shrimp fishery. The warm water, under controlled conditions, provides a favorable environment for maximum growth rate and the commercial outlook for this venture is promising. A similar application to catfish farming is gradually expanding, and it appears to have special growth potential in the Southeast.

SULFUR DIOXIDE

The criteria document on SO_2 published by the National Air Pollution Control Administration, states that most individuals will show a response when exposed to concentrations of 5 ppm (14,300 ug/m^3) and above, and that some sensitive individuals show detectable responses when exposed to 1 ppm (2,860 ug/m^3).[23] Note that reference is made to detectable responses, not physiological damage or acute distress. The work of Dr. Herbert Stokinger, Environmental Control Administration of the U.S. Public Health Service, points out that no health effects may be anticipated from sulfur dioxide, *per se*, at any conceivable community level.[24] This brings to mind

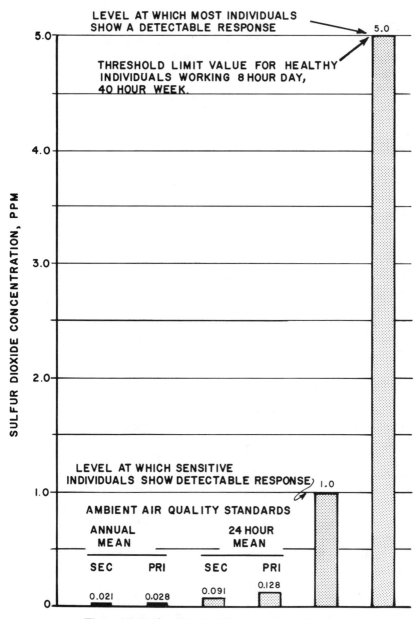

Figure 13. Sulfur Dioxide Concentration Values.

allegations of synergistic effects, and the work of Dr. Mary Amdur, Harvard School of Public Health,[25] and of Dr. Mario C. Battigelli, University of North Carolina School of Public Health,[26] make it clear that most of the allegations are completely unwarranted. Irritant response is increased by the presence of certain catalyzing aerosols made up of soluble salts of ferrous iron, manganese, and vanadium, but it is not increased by insoluble aerosols such as carbon, iron oxide, triphenylphosphate or fly ash. Even in those cases where there is an effect, the required concentrations of catalytic aerosols are greatly in excess of levels reported in urban air.[25] The American Conference of Governmental Industrial Hygienists established the threshold limit value for SO_2 at 5 parts per million for an exposure of healthy individuals working an eight-hour day, forty-hour week.[27] The only evidence to recommend limits as low as those adopted by EPA is based on correlations implicating SO_2 because of its presence, but cause-effect relationships have not been proven. In these cases, described in the criteria document already referenced, many other potentially causative ingredients were also present. A comparison of these different values for sulfur dioxide concentrations is shown in Figure 13.

There is another important point which seems to be overlooked in the trends in setting environmental standards. Sulfur compounds occur naturally in the atmosphere through decay of biological matter and from other natural sources. In fact, about two-thirds of the sulfur discharged to the atmosphere comes from natural sources. Furthermore, some sulfur in the soil and the atmosphere is beneficial to plant and animal life.[28-30] One writer suggests that the displacement of coal and wood by gas and oil for domestic heating, especially in rural areas, may have contributed to sulfur deficiencies in crop production because the substitution of fuels resulted in a net decrease in SO_2 in the environment.[29]

POLLUTION ABATEMENT OPTIONS

Particulate matter is generally abated by devices called electrostatic precipitators. In these devices, the dust in the flue gas is passed between electrodes, which give to these dust particles an electrical charge, causing them to migrate to a point where they may be collected and removed for ultimate use or disposal. Using the best of today's technology, it is possible to remove 99% or more of these particles, and the electric utility industry relies heavily on electrostatic precipitators for particulate matter abatement at the present time. The chief problems in this technology are that costs escalate rapidly as removal efficiency increases (see Figure 14), and that high removal efficiency is very difficult to achieve routinely. Efficiency requirements to achieve particular emission goals are shown on Figure 15.

The emission limit for particulate matter in a new power plant is 0.1 pounds per million Btu heat input. Achieving this limit will require removal efficiency of 99% or better, which strains the best of performance from current technology with electrostatic precipitators for coal-fired plants. Regardless of claims to the contrary, most industry experience leads to the conclusion that such performance can be achieved only when equipment is new and operating under nearly ideal conditions. There is a strong justification for cost/benefit analysis concerning this regulation since equipment costs increase sharply in the required range of performance.

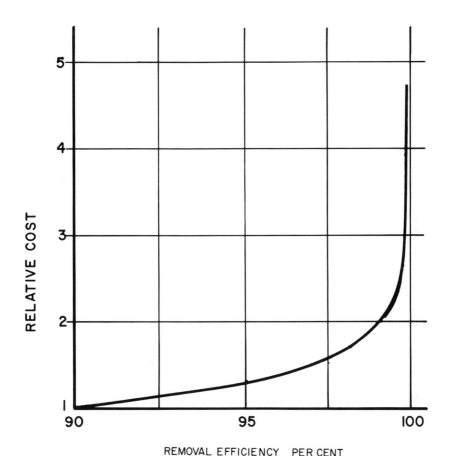

REMOVAL EFFICIENCY   PER CENT

Ref: ASME Standards No. APS–1

Figure 14. Removal Efficiency vs. Cost.

Other technology for particulate matter abatement includes bag filters and wet scrubbers. If wet scrubbing becomes feasible as a technique for SO_2 abatement, it will be fully explored for particulate removal as well.

SO_2 represents a different kind of problem. In existing plants, SO_2 constraints are stated, in most states, in terms of maximum ground-level concentrations that can be encountered downwind from the plant. This has required development of an understanding of the effects of local meteorology and natural atmospheric phenomena to bring about diffusion, dilution, and reaction of the SO_2 to remove it and reduce its concentration before it can reach the ground. High discharge stacks have been constructed in recent years

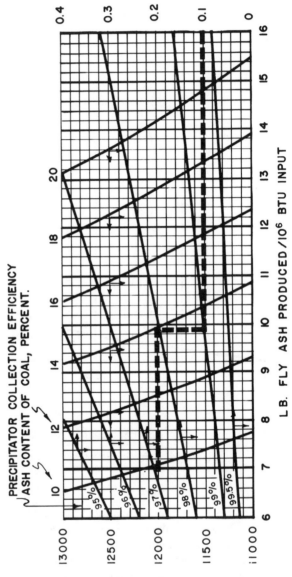

Figure 15. Collector Efficiency Required to Reach Emission Goals.

to provide this means of ground level control, and results have been very good.[33] Keep in mind that, through the use of high stacks, it is possible to control ground-level concentrations downwind from the plant and to comply with ambient air quality standards which require that concentrations averaged over a year should be less than 0.03 parts per million.

In federal regulations pertaining to new construction, imposed emissions limits such that the use of tall stacks is disqualified as an abatement strategy. This approach disregards the demonstrated effectiveness of tall stacks and denies the opportunity to utilize known relationships between weather regime and discharge height for achieving ambient air quality standards. On the other hand, the regulations, if implemented, will require installation of presently unproved technology which, if finally proven to be technically feasible, will reduce both plant reliability and plant efficiency. This will, in turn, require (a) the use of additional fuel resources, (b) the creation of more potential pollutants, and (c) the disposal of additional liquid and solid wastes not now associated with power generation. The related increase in both capital and operating costs for power generation will be appreciable.

The SO_2 emission standard for a new plant is 1.2 pounds per million Btu heat input. Sulfur content of most eastern coals ranges from approximately 1 to 5%. In Figure 16, an illustrative example shows that a coal with heating value of 12,500 Btu per pound and 4% sulfur content would require 84% sulfur removal either from the coal before combustion or from the flue gas after combustion in order to comply with this regulation. *There are two things which characterize this regulation: (a) it is unnecessarily restrictive if the purpose is to achieve the ambient air quality goals, and (b) there is no demonstrated technology available to enable compliance.* The industry is aware of the work to which EPA makes reference in connection with assertions that technology is available to achieve this standard. However, the EPA conclusions are not justified by the very limited experience afforded by the work in question.

Ambient air quality goals can be achieved by electrical generating stations without establishing emission limits. Accordingly the requirement of emission limits for power plants should only be used where ambient air quality standards cannot be achieved with strategies available to the plant. If ambient air quality standards cannot be achieved otherwise, then emission limits should be set so as to meet the ambient air standards, taking into account stack height, weather regime, and competing sources as mentioned later in connection with emission limits as an abatement strategy.

Air quality standards also exist for NO_x. However, the technological alternatives for abatement of oxides of nitrogen are either very limited or nonexistent. It appears that the best prospect for abatement of oxides of nitrogen is in the area of improved control of combustion temperature. In a gas-fired boiler, reductions in NO_x formation have resulted from limiting air additions in the primary combustion zone and delaying the addition of the remaining

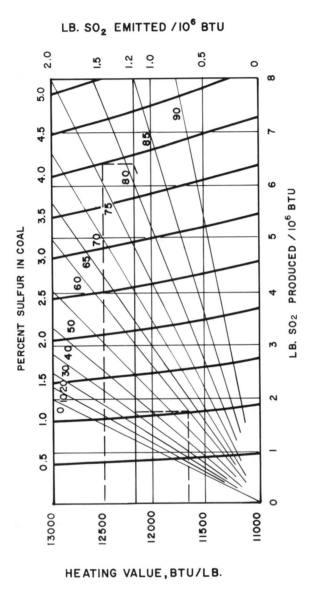

Figure 16. Example of New Plant Meeting Sulfur Emission Standard.

air needed for complete combustion to a point downstream. This is known as two-stage combustion, and it affects a reduction in NO_x formation by reducing the maximum temperature in the furnace. For an oil-fired boiler, a technique known as gas recirculation utilizes a portion of the flue gas to dilute and cool combustion air. This also has the effect of reducing the maximum furnace temperature.

Neither of these procedures has been demonstrated to be applicable to coal-fired boilers, and considerable work must yet be done before adequate technology can be available.[35] The production of NO_x is greater and abatement is inherently more difficult with coal than with gas or oil.

Considering the origin of particulate matter and SO_2 and the difficulty of removing these pollutants from the flue gas, it is obvious that technological avenues of abatement other than gas processing should be pursued. Toward that end, various organizations are investigating ways of processing coal to remove the ash and sulfur before combustion. One avenue is production of a low heating value gas from coal (not to be confused with the synthetic natural gas). It is expected that the production of low heating value gas can be achieved at lower cost than the production of synthetic natural gas. It would offer the same advantages as natural gas, such as freedom from ash and SO_2 problems, and it would be suitable for use in large gas turbines which are expected in the future.

Even more promising, perhaps, but further in the future, is the prospect of a fuel called "solvent refined coal."[36] In this process, coal is dissolved in a coal-derived solvent, and sulfur and mineral residue are removed. The solvent is then recovered for re-use, and solvent refined coal remains as the product. The fuel can be used either as a hot viscous liquid or as a solid in the form of flakes or little balls. Laboratory scale work was done on this process under sponsorship of the Office of Coal Research, an agency within the U.S. Department of the Interior. This process shows enough promise that the electric utility industry is sponsoring a development project to improve it and to expedite its commercialization. There is little doubt that coal refining of some kind offers an excellent growth prospect for it could, indeed, become as important in the country's future as petroleum refining is today.

While attending to the high-priority interests already mentioned, such as coal refining, breeder reactors, and the general concern about the environmental impact of power generation, the industry is also watching the developments of fuel cells, magnetohydrodynamic generation, fusion, and exotic ways to use the energy of nature such as heat from the sun, energy from the wind or ocean tides, or subterranean heat from hot springs. It is obvious to all that there are many problems ahead. Industry research efforts are intended to avoid some problems and reduce the magnitude of others and to provide information and direction for steady improvement in the required technology for meeting future needs.

EMISSION LIMITS AS AN ABATEMENT STRATEGY

The Environmental Protection Agency and many state agencies have elected to implement the achievement of ambient air quality standards by limiting the emissions from individual sources. If a fixed limit for emission is adopted, it follows that this limit must fit the worst case so that, when weather conditions provide the least dissipation of the pollutant, the ambient air quality standards can still be met. Thus, operating conditions for any emitter would have to be designed to constantly accommodate the conditions which prevail for only a few days or perhaps a few hours a year. This approach is in sharp contrast with a philosophy to use the assimilative capacity of the environment up to the point which the ambient air quality standards will permit. In order to use this philosophy, emission limits must be flexible, depending upon the local meteorological conditions, local discharge conditions, and locations of other emitters.

In the case of power plants, flexibility is reasonable, for several alternatives are available to each power plant for compliance with ambient air quality standards. For example, *for a given weather regime, emission limits could be defined in terms of stack height, with higher limits for higher stacks.*

Consider the points of contrast for these two different strategies for pollution abatement. In the one case, the strategy is to limit the emissions to a fixed value so that ambient air quality standards can be met at all times, but the assimilative capacity of the environment is not fully utilized except under the worst conditions. In the other case, where there is flexibility built into the mode of control, the power plant may use high stacks, fuel switching, reduced generation, or even fuel or flue gas processing (someday) to meet ambient air quality goals. This strategy would permit the maximum use of the environment while still safeguarding the ambient air quality and, hence, safeguarding the public interest. Quite obviously, application of such criteria to the process of setting emission limits is more difficult than setting a single value to be applied in all cases regardless of the need. On the other hand, it is wasteful of manpower and financial resources to work toward the achievement of an arbitrary fixed emission limit which reduces ambient concentrations below those needed to protect public health and welfare.

EVALUATIONS AND RECOMMENDATIONS

The missing ingredients in environmental regulations, as now formulated, are sensitivity to cost, sensitivity to available technology, and awareness of likely consequences, all of which are necessary for making rational cost/benefit decisions. Evaluation of each power plant and the regulations imposed on it should lead to achievement of reasonable protection of the environment at the least total cost to the public, where cost includes money, material (including fuel resources), manpower, related solid or liquid wastes, and environmental impact of alternative strategies (including an adequate electrical supply). In view of the obvious fact that resources are limited, it is appropriate to test various pollution abatement strategies in the context of conservation ethics.

1. Conservation of Limited, Irreplaceable Fuel Resources

Clearly, as efficiency increases, the fuel requirements decrease for a given electrical output. As noted previously, natural gas and oil reserves are meager in terms of national needs, and shortages are expected to worsen. Coal is relatively plentiful, but the use of coal is becoming increasingly difficult and expensive. Nuclear fuels are limited until and unless the breeder reactor technology is developed soon. Thus, every energy resource for power generation is destined to become increasingly more valuable to society, and the fuels being depleted cannot be replaced. Fuel requirements decrease as efficiency increases. Thus, heat rejection choices should maximize plant efficiency, and sulfur or particulate abatement choices should minimize "station service" and other irreversible energy losses.

2. Conservation of Land Resources

Each alternative for heat rejection has its own demands for land area, ranging from the minimum for once-through cooling up to 1 to 3 acres per megawatt for cooling ponds or artificial cooling lakes. Land value and land-use strategy should obviously be a factor in choosing the heat rejection alternative to be employed at any particular site. Whatever the choice, the total land area dedicated to heat rejection, fuel handling and storage, flue gas or fuel processing, or water disposal is decreased as efficiency is increased.

3. Conservation of Water Resources

Until recent years, water was regarded as a free resource. This is now an outmoded concept, and some authorities feel that water conservation may become the most serious of all constraints on power operation. It is obvious that fresh water supplies are limited, that there are numerous competing demands of society for fresh water, and that evaporative losses should be kept to a minimum. In every heat rejection strategy available today, evaporative losses are present in varying degrees. The heat rejection strategy at any particular location should appropriately account for the relative priority assigned to water resources. For a given choice, it is clear that increasing plant efficiency means a direct reduction in evaporative losses of water. Flue gas processing also has special water requirements which must be analyzed with care. In these processes, the analysis must include not only water losses but also the problems of contamination with chemical materials in solution which cannot be released to public waters.

4. Conservation of Financial Resources

Even while recognizing all of the many abilities and achievements of society, it must also be recognized that financial resources are not unlimited. The relevance of this statement is clear in connection with the costs associated with each of the heat rejection alternatives. Once-through cooling requires the least capital cost and the least operating cost and yields the highest plant efficiency of the heat rejection alternatives discussed. Any decision to abandon once-through cooling in favor of cooling towers, for example, would require an incremental increase in capital and operating costs and a decrease in plant efficiency. Similarly, arbitrary requirements for low-sulfur fuel, flue-gas processing, or any other strategy for achieving a particular environmental goal, will have associated costs. These costs, which must be borne ultimately by the society being served, are appreciable and society cannot afford such costs unless there are commensurate benefits. Up to this time, neither the costs nor the benefits have been adequately evaluated.

5. Conservation of Human Manpower Resources

Higher plant efficiency means lower power cost. Conversely, deliberate selection of a heat rejection strategy which lowers efficiency or a low-sulfur fuel strategy or a flue-gas processing strategy will increase power costs. The impact on the cost of manufacturing goods and of providing various services is sometimes subtle and indirect, but the direction of the influence is clear. Higher energy costs mean higher costs for goods and services to everyone, tougher competition with foreign goods both at home and abroad, and, in some cases, the difference between success and failure of a business venture with the resulting impact on employment. The point is that the incremental differences in the cost of implementing pollution abatement alternatives are large, that the consequences are important to society, and that a great deal is at stake in the long-term consequences.

OUTLOOK

As you review the scope of these comments, you will note there is no conclusion that there is an energy crisis, but it should be clear that there may be one. Society includes many forces at work which have a direct bearing on the question of whether or not there will be an energy crisis. It is imperative that great care be exercised at the federal level in formulating regulations which will govern the operations of power plants. The federal government is involved in the regulations which govern fuel procurement, mining operations, transportation, import and export quotas for fuels, air and water quality regulations, etc., and a great deal depends on the wisdom provided at the federal level.

The problems of nuclear power can be overcome. Construction costs and construction time can be brought under control, and continued attention to quality control can enhance public confidence in nuclear plants. The electric utility industry will buy and build nuclear power plants almost as rapidly as the nuclear industry can provide them. Because of various problems which are evident at this time, the estimate that nuclear will represent 25% of total capacity by 1980 seems optimistic. On the other hand, the estimate of 50% by the year 2000 seems to be attainable and may be conservative. As far as coal is concerned, there seems to be little doubt that processes for desulfurizing flue gas will emerge during the next five years and some large units will be successfully operated. Perhaps, beginning within five years and surely within ten years, commercial scale processes for making clean fuels from coal will be available and will supplant flue gas technology. This will lead to new types of petrochemical-electric utility interfaces that have not been contemplated before. Meanwhile, work will proceed in the development of unconventional power cycles and exotic concepts including, of course, breeder reactors and fusion projects. These developments will certainly be necessary in the latter part of this century in order to conserve coal and uranium for future generations. Research will probe perturbations in the thermodynamic cycles on which generating facilities are based, including media other than water and combinations of different media or different cycles for optimizing a mixture of output objectives. The concept of nuclear power plants as energy centers will surely gain support. More and more uses for energy will be converted to electric energy because of pollution abatement pressures and the recognition of electric energy as a convenient and clean energy form at the point of use. Electric mass transportation systems and electric personal vehicles will emerge as important elements in a future society.

REFERENCES

1. Price, Daniel O., ed. *The 99th Hour, The Population Crisis in the United States* (Chapel Hill: University of North Carolina Press, 1967).
2. Mayer, Lawrence A. "U.S. Population Growth: Would Slower be Better?" Fortune (June, 1970), 80–83, 164, 166, 168.
3. "Where Americans Will Live in 1970," U.S. News and World Report (April 10, 1972), 36–37.
4. "The Projected Growth in the Use of Electric Power," Chapter III, in *National Power Survey*, Federal Power Commission, Bureau of Power, 1970 edition.
5. Felix, Fremont. "Annual Growth Rate on Downward Trend," Electrical World (July 6, 1970), 30–34.
6. *United States Energy*, U.S. Department of the Interior, January, 1972.
7. Averitt, Paul, and M. Devereaux Carter. *Selected Sources of Information on United States and World Energy Resources: An Annotated Bibliography*, Geological Survey Circular 641, United States Department of the Interior, Washington, D.C.

8. Landsberg, H. H., L. L. Fischman, and J. L. Fisher. *Resources in Ameri-. ca's Future: Patterns of Requirements and Availabilities 1960–2000* (Baltimore: The Johns Hopkins Press, 1963).

9. "Another Area Where U.S.-Soviet Rivalry is Heating Up," U.S. News and World Report (April 10, 1972), 80–83.

10. Akins, James. "Can We Depend on 'Cheap Foreign Oil'?" The Conference Board RECORD (July, 1972), 23.

11. Wright, J. H., and John A. Nutant. "Energy Use and the Environment," Atomic Power Digest (4th Quarter, 1970), 2–5, Westinghouse Nuclear Energy Systems.

12. Tsivoglou, E. C. "Nuclear Power: The Social Conflict," Environmental Science and Technology (May, 1971), 404–410.

13. Corbit, C. D. "Reactors, Radiation, and Risk," Power Engineering (July, 1971), 43–45.

14. Jaske, R. T. "A Future for Once-Through Cooling," Power Engineering (February, 1971), 38–41.

15. "Dry Cooling Tower Shows Promise," Electrical World (June 1, 1972), 60–61.

16. Woodson, Riley D. *Cooling Towers for Large Steam-Electric Generating Units*. Paper presented at the Conference on Thermal Considerations in the Production of Electric Power, Atomic Industrial Forum, Washington, D.C., June, 1970.

17. "Wealth of New Data on Nuclear Plant Cooling Methods, Costs," Nuclear Industry (June, 1970), 7–15.

18. Woodson, Riley D. "Cooling Towers," Scientific American (May, 1971), 70–78.

19. Lee, William S. *Considerations in Translating Environmental Concerns into Power Plant Design and Operation*. Paper presented at the Conference on Thermal Considerations in the Production of Electric Power, Atomic Industrial Forum, Washington, D.C., June, 1970.

20. Kolflat, Tor. *Natural Body Cooling Water Techniques*. Paper presented at the Conference on Thermal Considerations in the Production of Electric Power, Atomic Industrial Forum, Washington, D.C., June, 1970.

21. Hauser, L. G., and K. A. Oleson. "Comparison of Evaporative Losses in Various Condenser Cooling Water Systems," *Proceedings of the American Power Conference, Volume 32*, 1970.

22. Warren, Frederick H. *Electric Power and Thermal Output in the Next Two Decades*. Paper presented at the conference on Thermal Considerations in the Production of Electric Power, Atomic Industrial Forum, Washington, D.C., June, 1970.

23. *Air Quality Criteria for Sulfur Oxides*, AP-50, National Air Pollution Control Administration, January, 1969, 156.

24. Stokinger, H. E. "The Spectre of Today's Environmental Pollution—

USA Brand: New Perspectives from an Old Scout," American Industrial Hygiene Association Journal (May–June, 1969), 208.

25. Amdur, Mary O. "Toxicological Appraisal of Particulate Matter, Oxides of Sulfur, and Sulfuric Acid," Journal of the Air Pollution Control Association *19* (9), 638–644 (September, 1969).

26. Battigelli, Mario C. "Evidence of Health Damage by Sulfur Dioxide," *Proceedings–Seventh Conference on Air Pollution Control*, Purdue University, October, 1968.

27. "Community Air Quality Guides," American Industrial Hygiene Association Journal (March–April, 1970), 255.

28. Robinson, E. and R. C. Robbins. *Sources, Abundance and Fate of Gaseous Atmospheric Pollutants*, Stanford Research Institute, SRI Project PR-6755, February, 1968, 26.

29. Coleman, R. "The Importance of Sulfur as a Plant Nutrient in World Crop Production," Soil Science *101*, (4), 230–238 (1966).

30. Ensminger, L. E. "Sulfur in Relation to Soil Fertility," *Bulletin 312*, Agricultural Experiment Station, Auburn, Alabama, June, 1958.

31. Clayton, J. Wesley. *Biological Effects of Sulfur Dioxide and Fly Ash*, Hazleton Laboratories, Inc. Paper presented at meeting of Edison Electric Institute, Boston, Massachusetts, June 2, 1970.

32. Wilson, W. E., and Arthur Levy. "A Study of Sulfur Dioxide in Photochemical Smog," Journal of the Air Pollution Control Association *20* (6), 385–390 (June, 1970).

33. Ross, F. Fraser. "What Sulfur Dioxide Problems?" Combustion (August, 1971), 6–11.

34. *Abatement of Sulfur Oxide Emissions from Stationary Combustion Sources*, Division of Engineering, National Research Council, Washington, D.C., 1970.

35. James, D. W. "Coping with NO_x: A Growing Problem," Electrical World (February 1, 1971), 44–47.

36. Hadley, R. C. "Solvent Processing Upgrades Coal," Electrical World (November 15, 1970), 58–59.

37. Greenspan, Alan. "Fiscal Constituencies," Barron's (October 12, 1970), 1, 8, 10.

8.
REMOVAL OF SULFUR OXIDES FROM STACK GAS: RE-COVERY VERSUS THROWAWAY METHODS

A. V. Slack

Tennessee Valley Authority
Muscle Shoals, Alabama

Of all the gaseous pollutants emitted by industrial operations, sulfur oxides have two distinctions: they are useful materials if recovered, and there is some chance of recovery being justified on an economic basis. In most cases, this cannot be said of other atmospheric pollutants.

There is already a considerable history of sulfur oxide recovery. One of the earlier examples is in Tennessee, at Copperhill. Smelters that treat sulfide ores emit large quantities of sulfur oxides, and, in many cases, there is the fortunate circumstance that the sulfur dioxide content of the gas is high enough to make direct conversion to sulfuric acid feasible. Thus, all over the world, there are smelters producing two products from metal sulfide ores—the metal and sulfuric acid.

The sulfuric acid industry has also adopted recovery in several plants. Here the situation is fortunate in another respect: although the sulfur oxide content of the tail gas is obviously too low for sulfuric acid plant feed and therefore must be concentrated, the concentrated stream from the recovery unit can be fed directly back into the acid plant, thereby avoiding the cost of new equipment to convert the sulfur dioxide to a finished material. Such "concentration" installations are in use in several countries including Japan, the United States, Czechoslovakia, and Romania.

The difference in situation between smelters and sulfuric acid plants points up one of the main problems in sulfur oxide recovery. For the acid plant, the main product is already sulfuric acid, for which there is presumably a market or the plant would not have been built. The recovery operation merely increases the output by a very small amount—on the order of 0.2 to 0.3%. In contrast, the smelter must enter into a new business if it recovers sulfur oxides and must take on all the problems associated with marketing a product for which the smelter is not ideally located and for which the company may not have an adequate background either in production or marketing. As a result, not all smelters recover sulfur oxides. The general practice has grown up of diluting the gas with air and emitting it from high stacks. The tallest stack in the world—1250 feet—is being built in Canada for this purpose. Such a stack is extremely expensive, so much so that the smelter company would

83

certainly recover the sulfur oxides if at all economically feasible. However, when the situation is such that there is some question whether the recovered acid can even be given away, some other course must be taken.

It is in this uncomfortable economic situation that the power industry now finds itself. Power plants fired with fossil fuels emit, in the United States, more sulfur oxide than all other industrial sources combined (Table 6), and the same situation holds generally throughout the industrialized areas of the world. The power industry, however, has none of the advantages enjoyed by smelters and sulfuric acid plants. The SO_2 concentration in the gas is so low that removal from the gas stream has not been given much serious consideration; until recently, emission through tall stacks was considered the best solution to the problem. And the sulfur dioxide obviously cannot be recycled as in sulfuric acid plants.

Table 6. Estimated Sulfur Dioxide Pollution in U.S. Without Abatement[a]

	Annual emission of sulfur dioxide (millions of tons)			
	1970	1980	1990	2000
Power plant operation (coal and oil)	20.0	41.1	62.0	94.5
Other combustion of coal	4.8	4.0	3.1	1.6
Combustion of petroleum products (excluding power plant oil)	3.4	3.9	4.3	5.1
Smelting of metallic ores	4.0	5.3	7.1	9.6
Petroleum refinery operation	2.4	4.0	6.5	10.5
Miscellaneous sources[b]	2.0	2.6	3.4	4.5
Total	36.6	60.9	86.4	125.8

[a]February 1970 estimates by the Air Pollution Control Office, excluding transportation.
[b]Includes coke processing, sulfuric acid plants, coal refuse banks, refuse incineration, and pulp and paper manufacturing.

Notwithstanding these problems, the current effort to improve the environment is rapidly pushing the power industry in the direction of taking steps to
reduce SO_2 emission. There are several possible ways of doing this, of which
four seem likely to be significant:

1. Use low-sulfur fuel. (This is discussed in another paper in this
   volume.)
2. Shift to nuclear power. The trend in this direction is strong, but the
   general consensus is that many fossil-fuel plants will be built in the
   future. Moreover, there is the problem of the existing coal- and oil-
   fired plants.
3. Recover sulfur oxides as a useful product and enter into the chemical
   marketplace.
4. Convert the sulfur oxides to solid form and dispose of in as harmless a
   way as possible.

For the near term, only 3 and 4 appear to be practicable. There is little
question that several of the processes now under development in these two
categories are technically feasible; only time and money are required to
develop the technology. The main question, then, is which of the two would
give the most practicable operating situation from the economic standpoint.

Recovery has received the most publicity. With the announcement of each
new process by its developer, the press has enthusiastically presented its
merits, even though in several cases the new departure has been only at the
"paper development" stage. Throwaway processes, on the other hand, have
not received nearly so much publicity, which is to be expected since they do
not represent the economic promise that everyone would like to see in this
waste material. In effect, the throwaway approach abandons the prospect of
converting a pollution problem into the attractive situation of recovering and
using a valuable national resource that is now being wasted.

There has also been widespread criticism of throwaway methods because
they merely convert atmospheric pollution to solid waste pollution. This has
been emphasized especially in urban areas where there is little space available
for the solid waste.

Nevertheless, the power industry is generally turning to the throwaway
route as the best solution to the problem, at least in the near future. TVA is
in process of installing limestone slurry scrubbing on a 500-mw boiler at the
Widows Creek station in northeast Alabama; the products, solid calcium sulfite ($CaSO_3$) and calcium sulfate ($CaSO_4$), will be ponded in the same way
that ash from the coal is now handled. Several other such installations are in
operation, under construction, or planned by the power industry in various
parts of the country.

Some recovery units are being installed on a full-boiler, commercial basis in
this country and in Japan. These are mainly test units, however, and
understood to be funded partially by the governments involved. It appears
that in all cases where the utility has selected a process for an internally

financed effort to meet pollution regulations, the choice has been a limestone throwaway process.

This does not indicate lack of interest in recovery or that the industry regards recovery as a hopeless approach. The consensus of those who have studied the situation thoroughly seems to be that the throwaway method is more fully developed, keeps the power company out of the unfamiliar chemical marketing field, and avoids the problem of forcing sulfur products into an already glutted market. Recovery processes need further development and evaluation, and must be introduced gradually as the market can be adjusted to accept the byproduct. The limestone processes, on the other hand, are relatively well developed and do not introduce marketing problems.

PROBLEMS IN THROWAWAY PROCESSES

The foregoing does not mean that the limestone methods are so well developed that design and operating problems are not expected. Actually, the technology is currently in a state of flux, with numerous test units (small scale, pilot, and full scale) in operation. Lime slurry scrubbing is used commercially in a sulfuric acid plant in Japan, an ore roasting operation in the USSR and in two power plants in the United States. The main problems that have been encountered are (1) boiler fouling (when limestone is injected into the boiler to calcine it to calcium oxide (CaO), a more reactive form), (2) scrubber scaling and corrosion, (3) erosion in the scrubber, (4) mist carryover, (5) cost of reheating the gas (to minimize loss of plume buoyancy), and (6) disposal of waste solids without polluting water courses.

The major process choice is between introduction of the limestone into the boiler and introduction directly into the scrubber (Figure 17). Boiler injection has the advantage: the resulting calcium oxide, carried into the scrubber with the gas, reacts more readily with the sulfur dioxide; thus a smaller and perhaps simpler type of scrubber can be used for the same degree of sulfur dioxide removal. Drawbacks are problems associated with dry grinding, the possibility of boiler fouling because the lime lowers the ash flow temperature in the boiler, increased dust load on the scrubber, and increased scaling in the scrubber.

Introduction into the scrubber gives the advantage of low-cost and pollution-free wet grinding, a less complicated situation in regard to retaining existing dust-collection equipment, and less scaling in the scrubber. The slower dissolution rate of limestone, however, requires more holdup in the scrubber, which can be attained by various means, all expensive. Moreover, there appears to be more danger of blinding reactive surfaces with reaction product when limestone is used.

Other considerations that are not yet adequately evaluated include scrubber type, type of limestone (mainly between calcitic limestone and dolomite), materials of construction, slurry circulation rate, solids content of slurry,

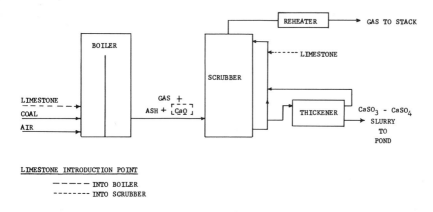

Figure 17. Removal of Sulfur Oxides From Stack Gas By Lime-Limestone Scrubbing; Alternate Points For Limestone Introduction Into the System Are Shown

stoichiometry, mist eliminator type, source of heat for reheating gas, and effect of magnesium content of the limestone on water pollution. Data on all these points will be obtained in the EPA-Bechtel-TVA test program to be carried out at TVA's Shawnee (Kentucky) station, equipment for which was installed in April, 1972. This intensive and extensive program is designed to test all aspects of lime-limestone scrubbing. Three different types of scrubbers are used in parallel, with introduction of the limestone both into the boiler and into the scrubber. Small-scale and pilot plant work is also under way at TVA and EPA and at various test units around the country sponsored by EPA, and power industry, and private engineering firms. Other countries active in current research are Germany, Sweden, Czechoslovakia, and the USSR.

One problem is the degree of sulfur dioxide removal needed. Almost any degree of removal can be obtained by lime-limestone scrubbing, but for the extremely low emission tolerance being considered (on the order of 100 ppm, or over 95% removal from a typical gas) it may be better to alter the process somewhat in order to reduce scrubber size and the slurry pumping rate required. The major possibility for this is scrubbing with an alkali salt solution (ammonium, sodium, or potassium) and regenerating the alkali outside the scrubber by treating the scrubber effluent solution with lime or limestone. The final product would be the same ($CaSO_3$ and $CaSO_4$), but the scrubber and pumping rate should be much smaller because of the high affinity the alkali salts have for sulfur dioxide as compared with lime or limestone. Small-scale work on this approach is being pursued at TVA, and plans are to do further testing in an "advanced concepts" pilot plant being constructed at the Colbert station near Muscle Shoals, Alabama.

The use of weak acids (*e.g.*, benzoic acid) in the scrubber circuit is another possibility for improving limestone slurry scrubbing. The acid is not consumed but goes through a series of reactions in the circuit that promote rapid dissolution of limestone. A small-scale EPA-TVA project on the method is under way.

An important aspect of lime-limestone scrubbing is that dust and sulfur dioxide removal can be combined with economic advantage. Wet scrubbers have been used very little for dust collection in the power industry in the past but will be incorporated in the lime-limestone scrubber systems now being installed. The main question is whether to use a separate scrubber stage (in series) for dust or to attempt removing both dust and sulfur dioxide in the same stage.

Full environmental protection requires that any water used to transport the waste solids to the disposal point be returned to the scrubber. Such a requirement brings in several more technical problems, all of which are under study. A major consideration is that soluble salts formed from constituents in the gas, ash, and limestone will build up the ionic concentration in the "closed loop" and possibly affect the scrubbing step.

It should be pointed out that, although smelters and sulfuric acid plants have a better situation in regard to recovery, these industries are also interested in the lime-limestone processes. This is particularly true for smelters located far from sulfuric acid markets, as many of them are. So far as is known, no smelters are currently planning use of lime-limestone scrubbing but future use seems likely.

STATUS OF RECOVERY METHODS

The growing emphasis on reducing sulfur oxide emission has produced a somewhat surprising number of recovery inventions. These new processes, added to the older ones that were developed mainly for smelters and sulfuric acid plants, make the budding technology so great in magnitude that it is difficult to even summarize it within the limits of this paper. There are some major categories, however, that can be listed as a means of gaining perspective on the subject.

1. Sorption processes in which sulfur oxides are concentrated as an intermediate material, followed by treatment of the intermediate to regenerate the sorbent and produce either a finished product or a rich stream of sulfur dioxide or hydrogen sulfide that must be treated further to get a finished product.

   a. Solid sorbents.

   b. Liquid absorbents.

2. Processes in which a finished product (either sulfuric acid or sulfur) is obtained within the main gas stream.

   a. Catalytic oxidation.

   b. Catalytic reduction.

The sorption processes outnumber all others by a wide margin. The earliest processes involved absorption of sulfur dioxide from smelter gas streams by liquids such as dimethylaniline and ammonia solutions. The main division in today's technology is between solid and liquid absorbents. There was some swing away from liquids in the 1940–1960 period because solids offered a way to avoid the problem of reheating the gas after the cooling caused by contact with aqueous solution in the scrubber. The many problems encountered with solids, however, have brought the emphasis back on aqueous scrubbing again.

Among the solid sorbents under study are activated carbon and the oxides of manganese, copper, and iron. In carbon adsorption processes, the sorbed sulfur dioxide oxidizes and forms sulfuric acid in the pores, from which it is removed by washing or by reducing to give a rich stream of sulfur dioxide or hydrogen sulfide. The metal sulfites and sulfates formed from reaction of the metal oxides with sulfur dioxide can be regenerated in various ways including ammoniation and reduction. The products in all cases are sulfur, sulfuric acid, or a salt such as ammonium sulfate. Both metal oxide (Mn_xO_y) and carbon are being tested on a commercial scale (100-200-mw) in Japan; no other plant-scale work is known to be in progress.

Aqueous absorbents include solutions of ammonium and sodium salts and slurries of metal oxides or hydroxides (calcium, magnesium, zinc). The ammonium and sodium systems are used commercially on smelters and sulfuric acid plants in the United States, Japan, Czechoslovakia, Romania, and the USSR. The magnesium oxide (MgO) slurry is being tested on a large scale in the United States and Germany. Regeneration procedure varies with process; sulfuric acid and fertilizer materials are the major products in the large-scale tests, and several ways of making sulfur from intermediates are being tested in small-scale work.

Catalytic oxidation in place to make sulfuric acid is the basis for one of the most advanced recovery methods; it is being tested on a large power plant boiler. Reduction to sulfur in place does not appear to have advanced beyond the bench scale.

COMPARATIVE ECONOMICS

TVA is carrying out for EPA a comprehensive series of conceptual design and cost studies on leading processes. It is considered that valid economic comparisons between processes cannot be made unless all the processes in question have been subjected to a conceptual design study and all the studies are made on a uniform basis. To date, studies have been made of the limestone processes, oxidation in place, and ammonia scrubbing (certain variations); the current study in progress is concerned with magnesium oxide scrubbing.

It is difficult to generalize on the results because comparisons depend so much on plant size, sulfur content of fuel, capacity factor, cost of limestone, new plant versus retrofit, and basis of financing. Investment estimating has been especially difficult because of the current rapid escalation in cost of building plants. As an order of magnitude, it can be said that lime-limestone scrubbing investment will probably be $30 per kilowatt or more, and that recovery will run somewhat higher. Operating cost for lime-limestone, assuming limestone delivered at $2 per ton, should range from $2 to $4 per ton of coal burned (0.75 to 1.5 mills/kwh).

Plant operating cost for recovery methods varies so widely with conditions (mainly with type of product) that it is useless to give a range. The main consideration in recovery, of course, is the net cost (or profit) after deducting return from sale of product. It is the difficulty in predicting the latter that gives power producers the greatest concern. The increasing flow of byproduct sulfuric acid from smelters and of byproduct sulfur from sour gas operations is likely to push the prices even further down from the current distress level.

Sulfur can be stockpiled if it cannot be sold at a reasonable price but sulfuric acid cannot, at least in a practical way. Possibly the best product is phosphate fertilizer made at or near the power plant, in which case the recovered sulfur would serve as a captive supply and would not have to enter the competitive market. This has been practiced to some extent by smelters, but they are not generally well located relative to fertilizer markets. Power plants would have a much better situation in this respect.

It is not expected that any recovery process can ever stand alone on its own economic merits for power plant use. The main question facing the industry is whether or not the net cost for controlling sulfur dioxide emission by the recovery route will be lower than for lime-limestone scrubbing. The EPA-TVA study on ammonia scrubbing (with production of phosphate fertilizer) indicated that recovery is cheaper if the plant is large (over 500 mw, to get large amount of product) and ideally located. However, only a few such plants would saturate the market.

CONCLUSIONS

Of the three major emitters of sulfur dioxide—power plants, sulfuric acid plants, and smelters—power plants emit the most sulfur dioxide but have several features that make sulfur dioxide recovery less promising than for the other sources. These disadvantages, added to the prospect of continuing depressed price levels for sulfur products, have led the utilities in the direction of adopting throwaway methods (absorption of sulfur dioxide by lime or limestone and discarding the product) to meet the growing restrictions on emission. Several such installations are in process or already operating. But recovery still has promise, under favorable situations and with further development, of being a less expensive way to control emission than the wasteful throwaway route. A large amount of development work on recovery methods is continuing both in this country and in other parts of the world.

9.
REDUCTION OF SULFUR OXIDE EMISSIONS BY FUEL MODIFICATION

Joseph W. Mullan

Vice President of Government Relations
National Coal Company

Fuel modification as a tool to reduce air pollution emissions is not new. Over the years many impurities have been cleaned from coal, *e.g.,* ash content. Today, sulfur oxide is the major concern and removal of sulfur from coal poses many totally different problems than those experienced in earlier coal cleaning operations.

Let me first discuss the forms of sulfur in coal, then review the removal of sulfur from coal prior to combustion and briefly discuss conversion of coal to gas.

Sulfur is universally present as an impurity in coal, not in elemental form, but as one or more of three compounds: (1) combined with the organic coal substances, (2) as an iron sulfide (FeS_2) in the form of pyrite or marcasite, and (3) combined as calcium and iron sulfates.

Sulfate sulfur in freshly mined coals generally is less than 0.05%. Sulfur occurring in the form of pyrite or marcasite is easily oxidized, however, especially under moist conditions, to the sulfate. The presence of sulfate in more than small amounts, therefore, indicates that the coal has weathered. Generally, sulfate sulfur content is always low and is not an important factor in coal utilization.

Pyritic and organic sulfur vary widely both in total amount and in the percentage which occurs in each form. For U.S. coals, total sulfur content varies from 0.5 to 6.0%, and organic sulfur may constitute from 40 to 80% of this total. Generally, organic sulfur predominates in low-sulfur coals. With higher total sulfur content, there is likely to be more of both the pyritic and organic forms, although there is no direct relationship between the two.

Not only does the amount and type of sulfur vary from place to place for a given seam, but it can also vary widely within a given mine.

Pyrites in coal may occur as substantial accumulations (as lenses and bands, joints or cleats, balls or nodules) or as finely disseminated particles. The size and distribution will affect substantially the amount of pyrite that can be removed by mechanical (physical) methods. Some pyrite can also be removed by chemical methods.

Organic sulfur is chemically bound to the coal in a complex manner, so that only drastic treatments sufficient to break the chemical bonds can separate the sulfur from the carbon and hydrogen in the coal. When the coal is subjected to such drastic treatment, the end product is usually something other than coal.

Mechanical methods of cleaning involves separation of the coal from waste products such as shale, pyrite, and roof-slate by utilizing differences in the physical properties—usually specific gravity—of these components.

Differences in the appearance of coal and shale and slate provided the basis for selective mining underground and for hand picking from moving belts. With increased mechanization of mining operations, selective mining became more difficult, the coal became more diluted with waste products, and hand picking became ineffective because of the smaller average particle size associated with mechanization. It is, however, frequently desirable to clean coal with a high ash content to reduce transportation charges and because the waste, primarily shale together with the sulfides contained in it, is a source of trouble, and so about 60% of all marketable coal is cleaned. Almost all of the cleaning was by mechanical methods.

The difference in specific gravity between coal and impurities forms the basis for 99% of all conventional mechanical cleaning processes for coal. The specific gravity of coal is about 1.3, of shale about 2.5, and of pyrites over 4.9.

Mechanical devices are divided into two general classes: wet (jigs, dense-media, concentrating tables, flotation) and pneumatic cleaning. Most wet and pneumatic methods involve a stratification effect. In jigging (48% of the total coal cleaned in 1965), the raw-coal feed is mechanically agitated in a fluid medium to form layers of material, with particles of lighter specific gravity (coal) on top and progressively denser material underneath. By an appropriate slicing mechanism, the layers are separated into "clean" product and refuse.

The dense-medium techniques (28% of total cleaned in 1965) operate on a float-and-sink principle; in a bath in which finely divided solids in water (or inorganic salts in solution) simulate a heavy-gravity liquid with a specific gravity greater than coal but lower than that of the shale, the "clean" coal floats on the surface and the heavier particles sink to the bottom. When fine coal (smaller than one-fourth inch) is cleaned in a dense-medium vessel, the gravitational process is aided by centrifugal action. Pneumatic processes (8% of total cleaned in 1965), used only for coal finer than about three-quarter inch, do not require drying of the coal after cleaning, but the efficiency of cleaning is generally low and is further lessened by variations in surface moisture content of the raw coal, frequently necessitating thermal drying of the coal prior to cleaning.

The type of cleaning method chosen depends on the size of the coal, the composition of the raw product, and the chemical-quality specifications imposed by the consumer. Two principal situations affecting to a consider-

able degree the choice of cleaning method prevail in the electric power industry: (1) where an existing plant is not committed to a single supplier on a
long-term contract; and (2) where a new plant with a variety of alternative
choices, including the choice of a long-term fuel supply contract, is being
considered. In the first instance, the consumer's choice of coal is limited by
the design of the plant. In the second instance, the final choice of a cleaning
method is more likely to depend on a combination of economic and environmental considerations.

Conventional mechanical coal cleaning methods, which in the past were
designed solely to lower the ash content of the coal product, are generally
effective in the removal of about 50% of pyritic sulfur from U.S. coals. Until
recently, sulfur reduction was not an objective in cleaning coal, and the pyrite
which is removed in the cleaning process is primarily that contained in the
coal matter as separate large particles.

Since pyrite is present as a discrete particle, it can theoretically be
separated mechanically from the associated coal. In most coals, however, the
pyrite particles are so small and so intimately mixed with the coal that
extremely fine crushing is necessary to break them apart and permit separation. As the particle size becomes smaller, the choice of cleaning method
becomes more restrictive and cleaning becomes more difficult and costly. In
addition to greater coal crushing costs, finer particles tend to retain more
moisture and invariably require expensive thermal drying before shipping.
The cost of the "clean" product depends also on the proportion of coal
which is lost in the waste fraction. *Fine crushing is normally not feasible at
the mine because the resultant fine coal powder cannot be shipped by conventional coal transportation means.*

Pyrite removal cannot be optimized unless the size distribution of the
pyrite particles is known; it is not possible to determine what reduction of
pyritic sulfur can be expected when coal is crushed to different top (maximum) sizes. Information on particle size of pyrites in various U.S. coal seams
is very limited, and some efforts are being made by Bituminous Coal Research, Inc. (BCR), and the Bureau of Mines to obtain additional information
with respect to the major U.S. seams. This is not an easy task since not only
does the amount and type of sulfur vary from place to place for a given seam
but it can vary widely within a given mine. For example, 37 samples from a
single mine in Green County, Pennsylvania, varied in total sulfur content,
proportions of sulfur forms, or removable pyrite sulfur.

Recognizing the need for additional research, the Bureau of Mines, BCR,
and others here and abroad have examined the following mechanical cleaning
processes: (1) electrostatic cleaning (nonconventional), (2) magnetic cleaning
(nonconventional), (3) froth flotation (very little used), (4) concentrating
table, and (5) air classification (nonconventional).

Electrostatic cleaning utilizes an electric field to produce differential movement of mineral grains resulting from differences in interfacial resistance to

the passage of electrons. Although this method eliminates the need for drying, the capital costs appear to be high. Furthermore, a large number of different variables affect the efficiency of separation. A relatively high loss of coal in the refuse seems to be inherent in this method. Total costs may exceed those required for removal of sulfur oxides from stack gases. However, application of this method to a stream in which the pyrites have already been somewhat concentrated in a modified grinding circuit, such as that proposed by BCR, might be economically attractive.

Magnetic cleaning utilizes the force of a magnetic field to produce differential movement of mineral grains resulting from differences of magnetic permeability.

Coal minerals are generally considered paramagnetic and therefore not amenable or very poorly amenable to magnetic separation. It appears that the magnetic susceptibility of pyritic materials can be enhanced if the coal is first heat-treated for short periods of time. Pretreatment, however, results in some devolatilization of the coal and loss of Btu value. Depending on many variables (such as amount and type of original sulfur; particle size; and temperature, duration and method of the heat pretreatment cycle), sulfur reduction by the magnetic process can vary from as little as 1% to nearly 80%. Much more research remains to be performed before commercial feasibility can be demonstrated.

Froth flotation utilizes differences in surface properties for the separation of impurities. Despite the large differences in weight between pyrite and coal, they have such similar surface properties that, under normal conditions, little separation can be accomplished by this method. The surface properties of pyrite can be changed, however, by the use of chemical additives. Still, it appears that, even under carefully controlled conditions, as much as 40 to 50% of the pyrite appears with the "clean" coal. Some sources believe that the only practical means to clean very fine mesh sizes of coal is by froth flotation.

Concentrating tables utilize differences in gravity for the separation of different minerals. In 1965, about 13% of the clean coal was upgraded by this method.

In many conventional coal washing plants in the United States, the coal is washed by a dense medium process, and the small sizes are treated on concentrating tables. While the primary objective generally has been to reduce the ash content, some removal of pyritic sulfur usually occurs. In some instances, between 60 and 70% of the pyritic sulfur has been removed in the combined cleaning operation. In some circumstances, the tailings may be sufficiently rich in pyrite to make a suitable feed for sulfuric acid manufacture. Some sources strongly recommend that this method be further explored, at least as a partial solution to the SO_2 pollution problem.

Air classification is a gravity separation method which has been actively researched by BCR. Since fine crushing is usually required to liberate pyrite from coal and since practically all modern power plants pulverize the coal to about the consistency of talcum powder before burning, BCR is exploring the possibility of incorporating a pyrite removal stage (air classification) between pulverization and combustion at the power plant. After investigating various techniques for cleaning the pulverized coal, BCR concluded that two or more stages would be necessary. In two-stage cleaning, about 50% of the coal, that which is completely pulverized in one pass through the mill, would be partially cleaned by an efficient air classifier, and the coarser material rejected by the classifier would be more thoroughly cleaned and then recycled. A number of factors, including the proportion of pyritic sulfur in the coal, will influence the effectiveness of this method in reduction of total sulfur.

In chemical methods of sulfur reduction, only pyritic sulfur is removed. With these methods, as with mechanical methods, pyrite particles must be physically separated from the coal substance before reactions for sulfur removal can be expected to take place at reasonable rates.

Two chemical methods—chemical oxidation and bacterial oxidation—essentially involve the conversion of pyritic sulfur to a water-soluble sulfate.

Chemical treatment techniques are very expensive compared with the procedures previously discussed, and it is doubtful that such exotic methods will advance to widespread commercial application. Because of the fine grinding required to expose the pyrite particles, chemical treatment would have to be conducted at the point of use to avoid increased transportation costs. Furthermore, the water used to remove the sulfates would have to be purified prior to disposal or re-use.

In addition to these various approaches to cleaning coal are those processes which convert coal to other forms.

For example, Spencer Chemical and Pittsburgh & Midway Coal Mining Co., working together, have performed original research sponsored by OCR on a process to produce a de-ashed, desulfurized fuel (some refer to this process as solvent refining). The process has been proven feasible on a range of coals, from high ranking to lignite. The end product would have a heating value of about 16,000 Btu per pound and could be handled readily by conventional railroad cars or be transported in a gas-solid or liquid-solid pipeline. Here, too, more research must be forthcoming before the project advances.

Technology for producing synthetic gas from coal has long been available. Commercial production of artificial illuminating gas was initiated about 150 years ago. This and other types of synthetic gas manufactured in the past, however, had a low calorific value, ranging from 120 to 180 Btu per cubic foot for producer gas to 425 to 575 Btu for carbureted water gas (as contrasted with an average of 1,035 Btu per cubic foot for natural gas). For this reason, long distance transmission of synthetic gas is uneconomical. The major emphasis in the development of coal gasification processes today is on the production of high-Btu gas with a minimum heating value of 950 Btu per cubic foot. A product of this quality could be blended with natural gas without seriously diminishing its heating value and could be transported economically through new or existing pipeline systems from points of manufacture to centers of consumption.

For different reasons, the federal government, coal interests, and elements of the natural gas industry have joined to support research and development in coal gasification: the federal government to broaden the energy resource base; the coal interests to develop new markets for coal; and the natural gas industry to insure a long range supply of economical gaseous fuel. Efforts directed toward coal gasification have increased significantly during the past five years.

The natural gas industry appears to be interested in the possibilities of using the coal reserves in this country to help supply the nation's future demand for gas fuel. This interest reflects some long range projections for the gas industry: (1) annual demand for gas may double in the next 25 years and discovery rates may not be able to keep pace with demand; (2) cumulative consumption of gas to the latter part of this century will therefore tax recoverable reserves; (3) imports from abroad—by pipeline from Mexico and Canada or as LNG from elsewhere in the world—are not expected to increase sufficiently to fill the gap between discovery and consumption rates.

While there are several reasons why coal is receiving favorable consideration, the real reason is that coal is an abundant indigenous resource. Recoverable reserves of coal and lignite, widely distributed across the country, are estimated at more than ten times those of natural gas, in terms of heating value. More than one-half of the reserves are estimated to be recoverable at not more than one and a half times present cost.

Gasification is the process in which coal reacts with oxygen, steam, hydrogen, carbon dioxide, or a mixture of these to produce a gaseous product suitable for pipeline transmission and subsequent use as a fuel. Gasification is an effective method of desulfurization because sulfur is readily removed and recovered as H_2S.

The four major processes for obtaining from coal a gas with heat content of 900 to 1000 Btu per cubic foot use variations of gasification-methanation. These processes, developed in cooperation with the Office of Coal Research of the Department of the Interior, are (1) hydrogasification, (2) CO_2 acceptor, (3) molten salt, and (4) two-stage superpressure. Much development is necessary if any of these four processes is to become commercially feasible in the next decade.

The hydrogasification process presently being demonstrated by the Institute of Gas Technology operates on a feed of pretreated coal (char), synthesis gas (carbon monoxide and hydrogen), and steam to convert more than 50% of the carbon in the char to gas. The most reactive portion of the char reacts with hydrogen to form methane, and part of the steam reacts with carbon in the char to form additional hydrogen. From the hydogasifier, gas goes to purification towers for removal of organic sulfur, hydrogen sulfide, and carbon dioxide. In a methanation step, remaining carbon monoxide combines with hydrogen to form methane and raise the gas product to pipeline quality.

The CO_2 Acceptor Process (Consolidation Coal Co.) is based on use of dolomitic or high-calcium lime for heat transfer and for removing carbon dioxide from the product gas. Process steps include: (1) crushing and drying raw lignite; (2) devolatilization and subsequent gasification of the lignite to produce a gas composed of methane, carbon dioxide (along with carbon monoxide), and hydrogen in the right proportions for pipeline gas; and (3) purification and catalytic methanation of product gas to yield high-Btu pipeline gas.

The molten salt process such as sodium carbonate supplies reaction heat and acts as a catalyst for the gasification reaction. The gasifier is divided into two sections by a vertical partition. The partition is perforated below the surface level of the molten salt so that the salt can circulate but the gas evolved on one side cannot be carried over to the other. The coal and steam enter on one side of the partition, and preheated air enters on the other. The coal residue carried through the partition by the molten salt is oxidized by the air to supply heat for the gasification reaction taking place in the other half of the reaction vessel. The gasification reaction is further enhanced by the catalytic properties of the molten salt, which lowers the required reaction temperature and optimizes methane formation.

In the two-stage superpressure process (BCR), fresh coal is introduced into a hot stream of synthesis gas where it reacts at an elevated temperature and pressure to produce high yields of methane directly from the volatile portion of the coal. The unreacted coal is swept out of the reactor together with the product gas. The solids are removed from the gas stream and routed to the lower section of the gasifier for use in generating the initial hot stream of synthesis gas by complete conversion under slagging conditions with added oxygen and steam. The molten slag is withdrawn and discarded. The raw product gas is cleaned and purified by removal of the carbon dioxide and hydrogen sulfide. The clean gas is finally converted to pipeline-quality gas by catalytic methanation.

I have chosen not to review each of these in depth because, quite frankly, they collectively represent the basis for a major segment of a future conference and the subsequent publishing of the papers presented at such conference.

CONCLUSIONS

None of the cleaning processes discussed in this paper results in total removal of sulfur. Mechanical and chemical methods of cleaning coal will, at best, result in only partial removal of pyritic sulfur—itself only a fraction of the total sulfur in the coal. In view of recent forecasts of the potential cost of removing sulfur from the flue gases or during combustion, elaborate cleaning or treating methods prior to combustion have little economic potential in the electric power industry except in the case of some established plants where addition of stack devices may not be feasible. Generally, coal preparation prior to combustion would make sense only where the application of a relatively simple and inexpensive method of cleaning would remove sufficient sulfur from the coal to make it an acceptable power plant fuel under terms of local restrictions. Under some conditions, it may be advantageous to use simple pretreatment of coal prior to combustion, in conjunction with a follow-up process such as limestone or dolomite injection.

Coal gasification does offer the future fuel user a "clean fuel"—no SO_2, no particulate, and reduced NO_x. But, again, the real stumbling block is time and money: time to demonstrate the various processes, and money to pay for them. The future of gasification appears to lie in providing, not a replacement for natural gas, but a supplement, as the cost of finding and using natural gas reserves increases. As a long-range, supplementary source of low-sulfur fuel, this method has promise for the future. Pipeline transmission of gas is generally more economical than transmitting electricity, and the production of this sulfur-free fuel will allow generation of electricity closer to the highly populated areas.

10.
FUELS OF THE FUTURE FOR COMBUSTION PROCESSES

John W. Tieman

Bituminous Coal Research, Inc.
Monroeville, Pennsylvania

Throughout our history, we Americans have believed that our natural resources were just about unlimited in abundance. We have developed habits of
household economy, but never learned to translate these private habits into
public policies. For us, the chief challenge was not in the conservation of
limited resources but the exploitation of seemingly unlimited resources. This
attitude led us on to conquering the wilderness and advancing the frontier.
But after these goals were achieved, the steady increase in population and the
rise in material expectations began to take their toll. During this century, the
toll has climbed steeply; as the century approaches the three-quarter mark,
we are becoming more and more concerned. In the last 30 years, our population has increased more than 50%. Moreover, this has been the century of
petroleum products, the private automobile, the airplane, the ready availability of electrical energy, space travel, nuclear energy, and many other
developments.

The result has been an explosive growth in our exploration and utilization
of resources. An oft quoted statistic is that the United States has 6% of the
world's population and consumes 30% of its natural resources. The question
is how long the supplies will last if we continue to use them up at the present
rate. No one can answer this question, of course, but we should begin now to
assess the relative value of our natural resources in relation to their abundance
and projected use in the future. In addition, the effect on our environment
should be re-evaluated with relation to our ability to comply to increasingly
stringent regulations on air and water quality.

The growth in energy demand in the United States has been almost un-
believable. There is usually some difference in opinion between various projections as to the exact amount of energy that will be required in the year
2000, but all agree that the demand will be of the order of three or four times
the 1960 level. To provide the fuel to enable this increase in energy consumption will be an enormous undertaking.

There appears, however, to be conflicting forces at work in our attempt to
satisfy this terrific energy demand. Take, for example, the electric power
industry. The rapid growth of this industry alone has required a doubling of
output every ten years, yet recently there have been difficulties and delays in

99

gaining approval for the construction and operation of new power plants designed to provide the additional power required. Shortages of fuels that can meet the increasingly stringent air and water quality regulations have compounded the problem. If both the demand for power and the demand for environmental protection are to be met, extensive research and development with the expenditure of considerable time and money will be required.

What are the energy fuel resources of the United States?

As with the projections of energy demand, the determination of fuel resources cannot be exact. Numerous estimates have been made by others more qualified, and a summary of their findings shows that the coal reserves of the United States so overshadow the reserves of our other fuel resources that the accuracy of the estimates is not critical. It appears that the probable minimum coal resources, at 1200 billion tons, are more than twice the probable minimum recoverable reserves of all other fuels combined.

Thus, it would seem logical that any change in our energy fuel use should be based on coal, our most abundant fuel. But people do not always take the logical approach to the solution of a problem.

One thing has not changed in the past several years. Setting aside atomic energy for the moment, no new fuel technology has been brought to the commercial stage for electric utilities. There have been a few modifications, but no real innovations. We still burn fuel to make steam to spin turbines to make electricity and, in the process, discharge heat and other waste products to the environment. That's exactly how it all started, and, if the process has been considerably refined, it has not been basically changed. Even now, it is not very efficient. It has produced cheap power, though, because fuel has been cheap.

Now we need new technology for the production of electricity to meet the growing demand for this form of energy while, at the same time, providing the required protection of the environment.

The ideal electric utility fuel should be reliable, abundant, inexpensive, and clean. Coal is reliable, far more abundant than any other fuel, and if it is no longer so low in cost as it once was, neither is anything else. As for cleanliness, new fuel technology will result in clean coal combustion or in control of pollutants after combustion. But, you may ask, what of the other fuels?

Domestic oil production cannot meet demand now, and has not for several years. We already import a quarter of our oil supply, including much of the oil used by the electric utilities. Under the pressure of air pollution control regulations, these shipments have been increasing to what many people believe is a dangerous level. The problem, as you know, is that Caribbean fields have about reached their limit, and new oil supplies lie mainly in the Middle East. Governments in that region are not noted for their stability, their good will toward America, or their unconcern toward profits.

Furthermore, a nation which is grappling with increasing deficits in its balance of payments as we are simply cannot afford for long to commit a rising share of its energy bill to foreign suppliers. Therefore, to protect both the national security and the national economy, it seems apparent that an alternative to Middle East oil must be found as soon as possible.

Atomic energy still has most of its old problems and has lately acquired new ones. The AEC is attempting to find a way around the recent court decisions delaying the operation of plants until their environmental impact can be evaluated. A legislative solution may be possible there, but no such simple answer is yet apparent for the problems of emergency core cooling. More fundamental, however, is the as yet ignored future problem which the coal industry has been pointing out for five or six years—the shortage of low-cost uranium to fuel light water reactors. Within the last year, the administration has acknowledged this by pushing ahead research on the fast breeder reactor, which will make more efficient use of our limited uranium supplies, but many problems remain before even the first demonstration plant is operating about 1980. If that AEC timetable is met, and the plant is successful, the utilities can begin ordering full-scale plants—but considering the delays encountered with light water reactors, it is certain that there will not be enough fast breeders on line to make a significant contribution to our supply of electricity before about 1990.

The AEC hopes that the first commercial breeder, the oxide-fueled liquid metal fast breeder, will have a doubling time of 15 years. But a breeder with a doubling time of 15 years just will not solve the problem of fissile material shortage because power requirements double faster than that. The breeder program will probably include a series of reactors until the ultimate, the "advanced carbide LMFBR," is perfected. In that unit, the doubling time may be short enough, but this is not a near-term reality.

What about natural gas? In Washington, D.C., today, you cannot get gas service for a new home let alone a new power plant. The same is true in many other cities—supplies are so short that pipeline companies and distributors will not add new customers or even provide additional service to old ones.

The Institute of Gas Technology recently predicted a gas deficit of 11.3 trillion cubic feet by 1980, 22.5 trillion by 1990, and 35.7 trillion by 2000. These figures even include imports from Canada and Alaska, and LNG. All the evidence indicates that not enough gas remains to be discovered to restore the days of plentiful domestic supply.

Negotiations are under way to import large amounts of liquefied natural gas—LNG—from Algeria and other foreign sources. But LNG imports also compound the perils of our reliance on oil imports—they place that much more of our energy supply under foreign control, and they add that much more to our balance of payments deficit.

Other alternatives are pipeline gas from Alaska or from Canada. But even if the ecologists can be satisfied, the Alaskan supply is fairly limited in terms of future demand. And additional gas from Canada is uncertain—the National Energy Board has voted down the idea at least for the moment.

Synthetic gas from naphtha natural-gas liquids, and crude oil is an alternative being actively pushed; already some 14 different SNG projects have been announced or proposed. The supply of naphtha is limited and most of that available must be imported, or refined from imported crude, which brings us again to overdependence on imports.

The real long-term solution to the gas shortage is synthetic gas from coal and eventually from oil shale. Research on pipeline-quality gas from coal has been going forward on a modest scale for several years, and four processes are now in the pilot plant stage. One of these has been developed by the laboratory for which I work, Bituminous Coal Research, Inc., in Monroeville, Pennsylvania, the research affiliate of the National Coal Association. Consolidation Coal Co. is preparing to operate a pilot plant in South Dakota, and the Institute of Gas Technology plant in Chicago recently operated for the first time. The Bureau of Mines is constructing its own pilot plant at Bruceton, Pennsylvania. Some companies are not waiting for completion of this research; the problem of supply is so acute they are contracting for large plants using a partially proven but expensive German coal gasification process.

In short, the time calls for some major changes in fuel technology for the electric utility industry. In fact, if the electric utilities are to meet the power demands forecast for the 1980s, new fuel technologies based on coal use must not only be designed and tested but must also be brought to commercial stage, manufactured and installed—and for this complicated and tremendously expensive task, the years between now and the 1980s are a dangerously short time. Fortunately, there is a beginning—small, underfinanced, but a beginning. Several processes hold promise for the 80s. Let me review briefly what are considered to be the most promising prospects for power generation with coal as fuel used more efficiently and with minimal environmental degradation.

One of the major deterrents to coal use is its high sulfur content. With the advent of regulations requiring stack emissions of sulfur oxides equivalent to 0.7% sulfur in the coal now, and 0.3% in a few years, all hope of obtaining compliance with sulfur emission restrictions by coal-cleaning, either by conventional processes or improved processes, has been abandoned. Reducing the sulfur content of the high-sulfur coals of the midwest and Appalachian regions by mechanically cleaning to such levels is impossible.

But there is new fuel technology which can accomplish the objective of converting high-sulfur coal to a low-sulfur fuel prior to combustion. One method, among many others, is the production of low-sulfur liquids, somewhat like a residual oil, by processes being developed by Consolidation Coal Co. and Hydrocarbon Research, Inc. Another approach, which produces either a liquid or a solid low-sulfur fuel from coal, is being pursued by The Pittsburg & Midway Coal Mining Co. These approaches are being sponsored at least in part by the Office of Coal Research, and there are others which are privately funded. All offer the desired objective: a modified coal product, low in sulfur and usable as a boiler fuel.

Another solution to the sulfur problem is to clean up the flue gases after combustion. I am sure that everyone is aware of the research under way, some of it in large-scale demonstration plants exceeding 100 megawatts capacity. This approach will be expensive, but it at least offers one avenue of relief which will enable coal to be used in conventional power plants within the restrictions on sulfur emissions. While problems have been encountered in applying this technology on a reasonably large scale and no truly commercial applications have yet been made, continued effort will resolve the problems involved.

However, it is obvious that the final answer does not lie in desulfurizing coal prior to combustion and/or recovering sulfur oxides from the stack gases of conventional steam electric plants. It is not the advanced power technology needed in the 80s and thereafter. Sulfur oxide removal equipment not only adds to the expense of the plant but also is a heavy consumer of power itself, thus making the plant less efficient, not more. This well may be the technological advance of the 70s, for it will be necessary in order to produce ample power within the demands of clean air, but for the 80s we should have something better. We should look to those processes which so modify the conventional coal-to-electricity cycle that they not only remove the pollutants but improve the efficiency of generating power from coal. Such approaches depend, for the most part, on the conversion of coal to a clean gas, not at the mine, but at the power plant.

You have read a great deal lately about gas from coal, but most of what you have read is about pipeline quality or high-Btu gas. There is a deterrent to the use of synthetic pipeline gas from coal for power generation, and that is related to the location of coal resources.

Each coal-to-pipeline gas plant will require high production coal mines and extensive coal reserves: at least 6 million tons annually for 20 to 30 years. While some suitable uncommitted blocks of coal are available in the East, it is probable that most gasification plants will be located in the West. The cost of moving gas from such plants to power stations in the East will be substantial.

However, Appalachia and the Midwest still have great reserves of coal, even if not in blocks large enough to support a pipeline gas plant. Much of this coal is high in sulfur content, and, under the new air pollution rules, its future is doubtful without equipment to clean the stack gases or some means of removing the sulfur before combustion. Low-Btu gasification offers another means of using these coal reserves for power generation without pollution problems. We would still gasify the coal, but not at a mine with subsequent shipment of the gas. For the application of low-Btu gas, the likely pattern is that coal be hauled by conventional train or barge from mines to on-site gasification plants at electric generating stations. This eliminates the need for large committed blocks of coal adjacent to the plant and allows the use of scattered reserves which lie closer to eastern and midwestern markets.

How does low-BTU gas and the technology for production differ from that of pipeline quality gas? In the first place, its heat content is about 200 Btu per cubic foot instead of 900 plus. And secondly, the production technique, while similar to that for pipeline gas from coal, is somewhat simpler.

How would such a gas produced on-site at a power generating station be used? In the simplest application, it would provide a clean fuel for firing a conventional boiler. But more important is the potential this approach offers for new power plants of the future; we open the prospect of not just another boiler fuel but a more efficient power generation technology. This is possible in a combined cycle process in which low-Btu gas would be burned in a gas turbine which spins a generator. The hot exhaust gases would be passed through a boiler to make steam for a second generator. The combined cycle process, under active study by several companies, holds the promise of an efficiency as high as 60%, in contrast to the 38% for conventional generating stations and lower capital cost.

We in the coal industry are encouraged by the fact that the pending Office of Coal Research budget contains $3 million for investigation of low-Btu gas processes. We would be more encouraged if the amount were greater, but it is a beginning.

There are other examples of advanced technology which offer increased efficiency in power generation based on coal. One of these is fluidized-bed combustion, which has been under investigation by workers both here and abroad. The basic concept consists of burning coal in a dense-phase, fluidized-bed of inert material. The high degree of interest in this combustion process stems from the apparent economic and technical advantages offered.

Another prospect is magnetohydrodynamics—MHD—in which power is produced directly from a high-speed stream of combustion gases moving through a magnetic field at high temperatures. The Office of Coal Research recently released a favorable evaluation of a proposed MHD process using powdered char from coal and concluded it may save up to one-third of the fuel required by present generating methods. OCR also has awarded a contract to Avco Corp. for a more intensive study of the MHD process.

Still another prospect is the fuel cell based on coal. Westinghouse did some laboratory work on this under an OCR contract several years ago. Improved materials are needed to achieve lower cost, but the efficiency of the process is high.

These are some of the prospects for new technology to meet the challenge of the future. To make these promises come true will require time and money and vision—and quite a bit of corporate courage. The nature of research is such that we may spend quite a lot of money and find we have discovered a way not to do something. That's where the vision and courage come in—if you know how to do it, the process is only engineering.

To summarize then, the fuels of the future for the United States must include coal for the supply of gas, oil, and electric power, and government must encourage the research to make this possible.

Coal makes up four-fifths of the nation's energy reserves, and other fuels are running short; coal is the only domestic source of energy capable of meeting future U.S. energy needs. No matter whose predictions or projections you use, including even those of the Atomic Energy Commission, coal must play a very important role in satisfying the national energy demands of the future. The coal industry, the electric utility industry, the gas industry, and the legislatures must recognize this fact.

11.
CONTROL OF ATMOSPHERIC EMISSIONS FROM
INDUSTRIAL BOILERS

W. C. Wolfe

Manager, Design and Improvement Engineering
Power Generation Division
Babcock & Wilcox, Barberton, Ohio

Pollutants, in the form of particulates and gases, from industrial boilers having a heat input greater than 250 MKB have been limited by the Environmental Protection Agency (EPA) effective December 23, 1971, on units which started construction after August 17, 1971. These limits are:

Emission	Fuel	Lb/MKB Input	Approx. ppm (dry)
Particulate	All	0.1	(0.12 grains per SCF)
SO_2	Liquid	0.8	550
	Solid	1.2	520
NO_x	Gaseous	0.2	165
	Liquid	0.3	227
	Solid	0.7	525

Although recovery boilers for the pulp and paper industry are not now covered by EPA requirements, there are numerous state pollution requirements for emissions from these units. Methods to control atmospheric emissions in the past have been the primary responsibility of the equipment manufacturer. The new regulations present an even greater challenge.

Gaseous emissions occur when any fuel is combusted in a boiler furnace; particulate emissions occur when ash-bearing solid or liquid fuels are combusted. Objectionable gaseous emissions are sulfur dioxide, carbon monoxide, nitrogen oxide, and total reduced sulfur compounds (TRS) such as hydrogen sulfide, mercaptans, and other organic sulfides.

Methods of controlling emissions depend on the fuel being burned and the particular equipment used to burn the fuel. There are numerous types of fuels combusted in many variations of equipment. This discussion on controlling emissions will be limited to a typical boiler for general industrial use firing oil and gas as fuels, and a typical kraft recovery boiler utilizing kraft liquor as the fuel. The subject will be discussed by the type emission rather than the type

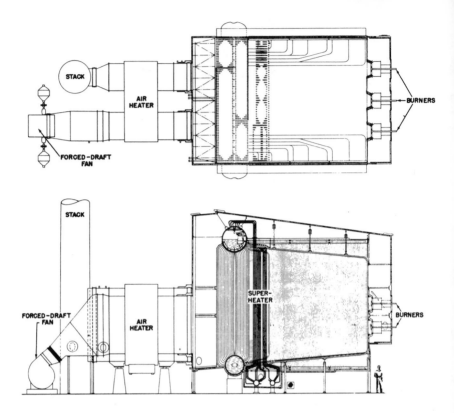

Figure 18. Field Erected Oil and Gas Fired Industrial Boiler

unit since, in most cases, the emissions are similar and the method of control varies.

Figure 18 shows a field erected industrial boiler rated at 360,000 lb/hr steam flow. There are six oil and gas burners on the front of the boiler where the fuel and air are mixed for combustion in the furnace. The flue gas flows over the generating bank in a three pass arrangement before going to the air heater and to the atmosphere. This type unit is used to generate steam for industrial purposes.

Figure 19 shows a field erected kraft recovery boiler rated at 492,000 lb/hr steam flow and 1000 tons of pulp. Black liquor at 60% solids concentration (40% water) is sprayed into the furnace through the front and rear walls. Some of the liquor burns in suspension, and the remainder partially dries on the walls and falls onto the furnace floor in a pile called a bed. Here, primary air is added in sufficient quantities to cause the liquor to smelt under a reducing atmosphere. The smelt flows from the furnace and is reused in the mill cycle. Additional air (secondary and tertiary) is added at higher eleva-

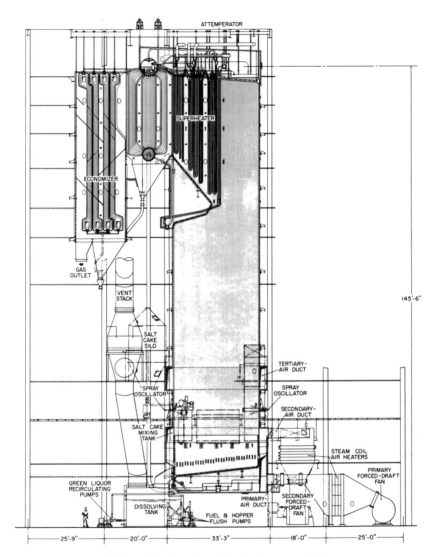

Figure 19. 1000-ton Kraft Recovery Unit

tions in the furnace to complete the combustion process. The flue gas then flows up the furnace into the superheater, generating bank, and economizer before entering a particulate collector, and then to the atmosphere. This type unit is used to process chemicals and generate steam.

PARTICULATE EMISSIONS

Particulate emissions can be controlled by use of (1) mechanical collector, (2) electrostatic precipitator, (3) wet scrubber, (4) boiler design, (5) operational procedures, and (6) selected fuels.

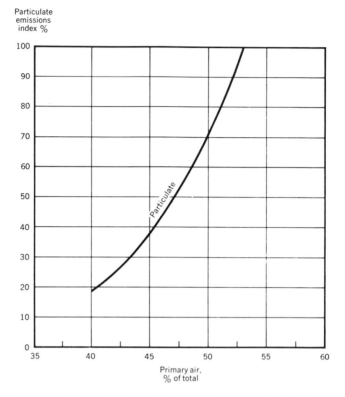

Figure 20. Effect of Primary Air on Recovery Boiler Particulate Emission

Selection of the collector equipment depends on the fuel fired and effi- ciencies required to collect the particulate matter. Factors affecting equip- ment selection are (1) dust loading, (2) grain size, (3) flue gas weight, (4) flue gas temperature, (5) chemical analysis of ash, and (6) flue gas composition.

Boiler design can reduce the amount of ash going to the collector, resulting in a reduction of stack emission for similar collector efficiencies. Liberal furnace design, good burner performance, and well-designed gas flues for improved gas and dust distribution are factors which reduce and distribute dust loadings to the collectors.

Proper atomization of fuel oil and turbulent mixing of the atomized oil particles and air will reduce the carbon content in the ash.

Proper black liquor and air distribution will reduce the carryover of ash in recovery furnaces. The proportion of air used at the three levels is important. The amount of air supplied to the bed is of particular importance. Figure 20 shows that the particulate matter from the bed increases with an increase in the percentage of primary air to the bed when other factors are held constant. Selection of low ash fuels will reduce the dust loading to the collector.

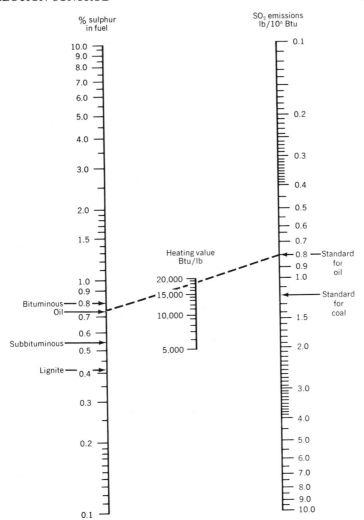

Figure 21. Nomograph Relating Sulfur in Fuels and Emissions

GASEOUS EMISSIONS

Carbon Monoxide

Low carbon monoxide levels can be obtained on oil and gas firing by burner designs which insure intimate mixing of air and fuel. On recovery boilers, good air and liquor distribution is a requirement to insure low levels of carbon monoxide in the flue gases.

EPA limits do not apply to carbon monoxide emissions, but many states have included them in their regulations.

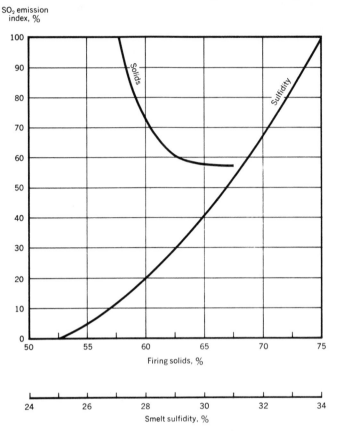

Figure 22. Effect of Fuel Characteristics on Recovery Boiler Emissions

Sulfur Dioxide

The amount of SO_2 emission on oil and gas firing depends on the *sulfur* content in the fuel. This can be calculated from the following formula and is shown in nomograph form on Figure 21:

$$\text{LB } SO_2/\text{MKB input} = \frac{.0198 \times S \times 10^6}{\text{BTU/LB}}$$

where S is the percent of sulfur in fuel
BTU/LB is the heating value of fuel

Emissions can be lowered by reducing the sulfur content of the fuel or installing an SO_2 clean-up system. At the present time, these systems have not been developed to the point where any are economical for industrial boilers so that reduction of fuel sulfur becomes the best method of SO_2 emission reduction.

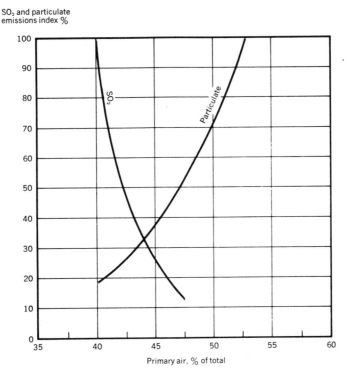

Figure 23. Effect of Primary Air on Recovery Boiler Emissions

The amount of SO_2 emission on recovery boilers depends on the black liquor fuel characteristics including solids, Btu, sulfidity, air temperature and distribution, and proper operating practices. Trends can be shown by holding all variables constant except one, and noting the effect on SO_2 emission. Figure 22 shows the relationship of two fuel variables where an increase in solids results in a decrease in SO_2 while an increase in sulfidity results in an increase in SO_2 emissions. Figure 23 shows that an increase in the percentage of primary air to the bed results in a decrease in SO_2 emission. Since we previously showed an increase in particulate carryover for the same condition, it is important to note that an optimum setting must be made to minimize both emissions.

Total Reduced Sulfur Compounds

While the other gaseous emissions are fairly well known to industry and the public, total reduced sulfur (TRS) compounds are primarily known in areas around a kraft paper mill. The "rotten egg" smell is primarily hydrogen sulfide.

Once again, proper fuel and air distribution is the key to low emissions. Elimination of the direct contact evaporator or use of oxidized liquor permits lower TRS emissions to the atmosphere. Proper operation of the recovery furnace can result in TRS levels below 1 ppm. Previously noted was the importance of the amount of primary air to SO_2 and particulate emissions. Proper use of secondary and tertiary air will reduce the TRS emissions. The amount of primary, secondary, and tertiary air for optimum emission control will vary for each unit, depending on the fuel characteristics.

Nitrogen Oxides

Methods of control on other gaseous emissions have been known for some time. More recently, attention has been directed to nitrogen oxides emissions. Most of the work on boilers has been on utility units. The EPA regulations include a large segment of industrial boilers, and manufacturers are presently developing various means for controlling these emissions. The known methods of controlling NO_x emissions are (1) two-stage combustion, (2) gas recirculation, (3) selection of furnace size, (4) selection of fuel oil, and (5) burner design. One or more of these methods can be used to lower the NO_x emission.

Burner design, combustion air temperature, excess air, gas recirculation, two-stage combustion, and furnace size affect the flame temperature of the burners and the thermal nitrogen oxide generated. The nitrogen content of the fuel (fuel nitrogen) affects the amount of nitrogen converted to nitrogen oxide in the flue gas. The work in several research facilities has shown that up to 50% of the nitrogen in liquid fuels can be converted to NO_x. The nitrogen content of No. 6 fuel oil varies between 0.1 and 0.6%; No. 2 fuel oil has much lower values, in the range of 0.01 to 0.02%. Figure 24 shows the predicted fuel NO_x in lb/MKB for various fuel nitrogens and conversion rates without emission control. To this value, the amount of thermal NO_x must be added. It is easy to see that using the lower nitrogen fuel goes a long way in reducing the NO_x emissions, but that another effective control means is still required.

Control of thermal NO_x with the means previously described has been successful on utility boilers. Use of two-stage combustion limiting the available air at the burners to 85% of the theoretical requirement has resulted in reductions in NO_x emissions of 75% on gas firing and 25% on oil firing. Gas recirculation mixed with the combustion air has also been proven a means of limiting NO_x formation. When used in conjunction with two-stage combustion, an additional reduction was obtained on gas and oil firing.

Laboratory work has shown that gas recirculation is effective in reducing NO_x emissions due to thermal fixation but has little or no effect on fuel nitrogen conversion. Two-stage combustion is effective in reducing NO_x emission due to thermal nitrogen and fuel nitrogen.

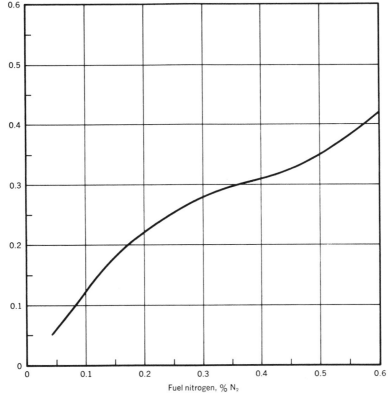

Figure 24. Effect of Fuel Nitrogen on NO_x Emission

Field work on an industrial boiler using two-stage combustion lowered NO_x emission by 40% on gas firing. Installation of NO_x ports in the furnace and the use of high static air as shown on Figure 25 provides a simple and economical method to control NO_x emission. Approximately 20% of the combustion air is introduced into the furnace through these ports, which results in the burners operating at approximately 90% of stoichiometric air in the first stage.

Field data shows a relationship between furnace liberation rate and NO_x emission. Increasing the amount of surface exposed to the fire results in a reduction of NO_x generated in the furnace.

Field data has also shown that highly turbulent burners and use of pre-heated air gives higher peak flame temperatures with an increase in thermal

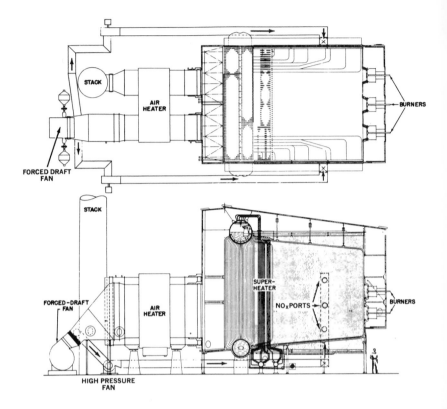

Figure 25. Field Erected Oil and Gas Fired Industrial Boiler with NO_x Ports

NO_x liberation. As previously noted, this type burner is desirable for low carbon monoxide and particulate emission. More development work will be required in this area in an effort to obtain low emission rates for all pollutants and still burn the fuel satisfactorily from a combustion standpoint.

SUMMARY

There are methods available to control atmospheric emissions from industrial boilers. The methods used depend on the fuels burned and the requirements to be met. Design of the equipment plays an important part in the control of emissions, as do operating practices. In some cases, a change in fuel chemistry will be required to lower the emissions to an acceptable value.

Section III

Metallurgical Processes and Air Pollution

12.
EVOLUTION OF FLUORIDE RECOVERY PROCESSES ALCOA SMELTERS

C . C. Cook

Technical Manager, Smelting Division

and

G. R. Swany

Chief Process Engineer, Smelting Pollution Control
Aluminum Company of America
Pittsburgh, Pennsylvania

EARLY HISTORY

The Aluminum Company of America first designed fluoride control into Niagara Falls Plant No. 3 in 1907. Control of fluoride was in the form of a three-story plant with the cell room located on the top floor so that cooling air could be drawn by natural draft through floor grilles located in the floor between each pair of cells. Air flow was upward through the building and out through four large, square ventilating towers on the roof. These towers extended almost 100 feet above the roof and created a strong natural draft. This rudimentary environmental control was primarily for protection of employees.

The design of potrooms, or cell rooms, was further improved in installations at Alcoa, Tennessee, in 1914; and at Badin, North Carolina, in 1916. Improved building designs provided natural circulation of air and exhaust of fumes through roof vents or monitors.

As early as the 1920s, investigations were made of plant surroundings to see if there was damage by escaping fluoride. There was no indication of damage at that time. It should be noted that pre-World War II aluminum plants were small by today's standards. One modern potline can produce almost as much metal per year as the entire plant at Alcoa, Tennessee, produced in 1941.

DEVELOPMENT OF CELL GAS COLLECTION AND TREATMENT FACILITIES

Alcoa first received indications of possible damage by fluorine from smelting plants in 1943 in connection with a Defense Plant Corporation plant located at Riverbank, California. At this location, trade winds blew in one direction for approximately 10 months of the year onto pasture fields downwind. The plant was not equipped with facilities for collecting and treating pot gases. A program was begun to design and install water sprays in the building monitors to wash the soluble gaseous components and precipitate the particulate matter from exiting air. However, before the work progressed beyond preliminary steps, the plant was shut down with the end of World War II.

The monitor spray development program was moved to the Vancouver (Washington) Works and a system was designed, installed, and tested. The system was 40 to 60% effective in removing fluorides escaping from the pots. However, after the equipment was installed in the plant, problems with building and equipment maintenance developed and work at Vancouver, Massena, New York, and Alcoa, Tennessee, continued toward a better way to reduce fluoride emissions.

By 1946, the smelting plant at Alcoa, Tennessee, was becoming concerned about possible damage to cattle in the area around the plant. Additional effort was applied to design and construct appropriate equipment which would, to the greatest extent practical, eliminate the emission of fluorides from smelting plants. The Company also instituted research so that toxic levels from ingestion of fluorides might be determined and diagnostic criteria of fluorosis scientifically ascertained.

There are two types of smelting cells: those using prebaked anodes, and those using Soderberg anodes. The prebake cell uses multiple carbon blocks which have been separately baked and attached to a conductor rod. The anode assemblies are replaced individually as the carbon is consumed. The Soderberg cell has a continuous anode contained in a rectangular compartment suspended over the cell. The anode materials, in the form of a paste, are introduced into the top of the compartment and are baked by the heat of the cell as they move downward in the compartment. Electrical connection is accomplished by either horizontal or vertical spikes embedded in the anode.

While Vancouver and Massena had only prebake anode cells, Alcoa, Tennessee, had both prebake and horizontal-spike Soderberg cells. Even though the monitor spray equipment was in use in Vancouver, it was believed that a more practical and efficient solution would be to enclose the individual smelting cells and collect the effluent which could then be pulled out through ducts to treatment equipment. Work began to determine the most practical method of enclosing the prebake anode cells and the volume of gas to be treated. This included investigation of chain curtains, asbestos curtains, and sheet metal covers as well as different volumes of air. The sheet metal hooding proved to be the most effective method of containing gases from prebake cells. Work was also done to find the best way to incorporate the fume-collecting duct in the existing cell structure.

The horizontal-spike Soderberg cells presented a more complicated problem. Attempts to enclose individual cells resulted in   numerous environmental and operating problems. The cells were enclosed in a "dog house" which effectively collected the gases. However, when the structure was opened to allow access to the cell, the heat and dust caused discomfort to the workers. After much experimentation, it was decided that the best solution would be to convert these potlines to vertical-spike anodes. This decision was expensive and the problems to be overcome were major. To make the conversion required (1) developing an anode which could be used with the existing cell cathode, (2) developing a means to enclose the cell and collect the gases, (3) determining proper gas volume, and (4) developing a system to treat the collected gases.

Development of gas treatment for both prebake and Soderberg cells continued with investigation of various methods of scrubbing the collected pot gases. One need which became apparent was a burner as part of the Soderberg hooding and collecting system to burn the hydrocarbons evolved from the anode to keep them out of the scrubbers.

Considerable time was devoted to developing the most practical method of scrubbing the gases. This included work with Pease-Anthony scrubbers, spray towers, Venturi scrubbers, and packed towers using glass, wool, coke balls, etc. The end result of this work was the determination that the simplest wet scrubber would capture most of the gaseous fluorides and that the most elaborate would not capture the finer particulate fluorides. The simple scrubber would achieve 80 to 82% total fluoride efficiency while the more elaborate devices would achieve 86 to 88% efficiency. Throughout the equipment development program, it was also necessary to develop sampling and analytical procedures to allow proper evaluation of the equipment being tested. Alcoa Research Laboratories worked with plant personnel and outside consulting firms to develop and evaluate procedures and methods.

The treatment system which evolved for prebake pots included a Rotoclone to move the pot gases and to remove the larger particulate matter and scrubber towers to remove gaseous fluoride and some of the remaining particulate. The larger particulate was collected and returned to the cells, but the gaseous components were lost in the water. Soderberg gas treatment equipment could not include Rotoclones due to the hydrocarbon problem.

Concurrent with the hooding and treatment development program, the Company had sponsored, either alone or with others, experiments to demonstrate the toxic level from the ingestion of fluorides by livestock and to determine the symptoms and other diagnostic criteria of fluorosis as distinguished from the general symptoms of other maladies.

Installation of the hooding equipment, ductwork, Rotoclones, and scrubbers on the Vancouver potlines was completed in 1949. Independent studies indicated that the system was effective in removing fluorides from the pot gases.

IMPROVING TREATMENT TECHNIQUES

Installation of remedial equipment was completed at Alcoa, Tennessee, by the end of 1952. In the interim, further development work had shown that electrostatic precipitators were more effective than Rotoclones in removing particulate from pot gases. New prebake plants in Texas and Washington state, which were designed and built in the late 40s and early 50s, had fume collection ducts designed into the cells and the fume treatment facilities employed electrostatic precipitators and scrubber towers (Figure 26). In addition, the design of the Texas plant included multiclones to collect additional particulate. A Soderberg plant in Texas was designed to incorporate fume collection equipment and scrubber towers.

In obtaining land for these new smelter locations in Texas, Washington, and later in Indiana, Alcoa purchased considerable acreage surrounding proposed plantsites. Additional property was obtained after the plants were put on stream. While the main purpose was to provide a buffer zone around the plant to insure no damage being incurred by our neighbors, the property has not become a wasteland. Studies were made to insure that land would be used for its most suitable purpose. Some of the best apples, pears, and peaches in the country are grown near the Wenatchee (Washington) smelter. Soybeans and corn are raised next to the Warrick (Indiana) operations. In addition, the company maintains a large prize beef herd, administered by the University of Tennessee, on land surrounding Tennessee Operations. The Badin (North Carolina) Works is nearly surrounded by a residential community directly adjacent to the plantsite.

At Alcoa, Tennessee, while believing that the small amount of fluorides still escaping from the plant could cause no damage to any of the surrounding farms, we decided to take the further step of installing electrostatic precipitators. The addition of dry precipitators on the prebake lines was completed in 1954, and addition of wet precipitators on the Soderberg lines was completed in 1955.

A lawsuit was filed against Alcoa in 1955 by a group of area farmers. The court ruled that there had been no damage from fluoride emissions after precipitator installation in August 1955.

DEVELOPMENT OF DRY PROCESSES

All of the fluoride treatment facilities installed prior to the late 50s had been "wet systems," but we were not content with results. Aluminum fluoride used in smelting is an expensive commodity, and it is desirable to recover the fluoride captured in a treatment system. In any event, the fluoride must be removed from scrubber water to prevent a water pollution problem.

Dilute hydrofluoric acid recovered in the scrubbers must be neutralized by treating with lime and allowing the resultant calcium fluoride sludge to settle out. The calcium fluoride, the form in which fluoride is mined as fluorspar,

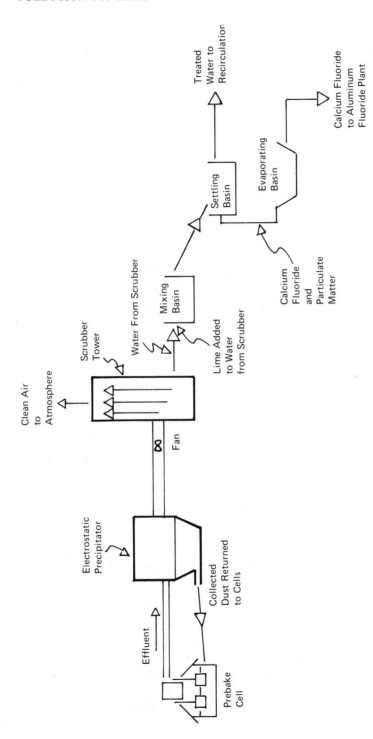

Figure 26. Typical Electrostatic Precipitator Plus Scrubber System.

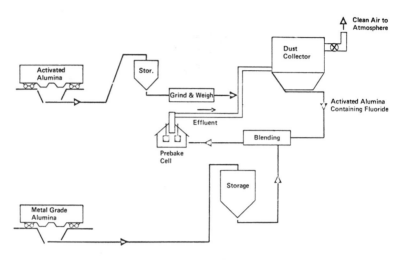

Figure 27. Dry Process.

can then be processed to make aluminum fluoride, which is returned to smelting cells. While most plants have water treatment facilities, it is not economical to construct aluminum fluoride production facilities at each location. The plants recover their calcium fluoride "sludge" and ship it to a central location, where it is converted to aluminum fluoride.

To eliminate water treatment and fluoride recovery processes, a dry system which will capture fluoride in usable form is most desirable. While installing the best known technology in "wet" scrubber systems in the early 50s, Alcoa pursued development of a dry process. By 1957, a process was available which captured the fluoride in a directly reusable form with gas treatment efficiency equal to, or better than, most wet systems (Figure 27). This system, however, had the disadvantage of requiring a special, extremely fine alumina. This required grinding equipment at the plant to obtain the desired alumina size and blending equipment to mix the special alumina after it had been through the fume treatment process, with metal grade alumina for use in the smelting cells. While installing the dry system on new potlines in the late 50s and early 60s, the search continued for a process to eliminate the grinding and blending problems.

THE ALCOA-398 PROCESS

Before 1957, it was known that HF gas in small quantities would react with alumina at low temperatures. Experiments indicated that an apparatus could be installed on a smelting cell to capture the pot gases, strip them of fluoride and particulates, and return the recovered fluoride to the cell with the alumina feed. However, plant scale tests resulted in equipment problems.

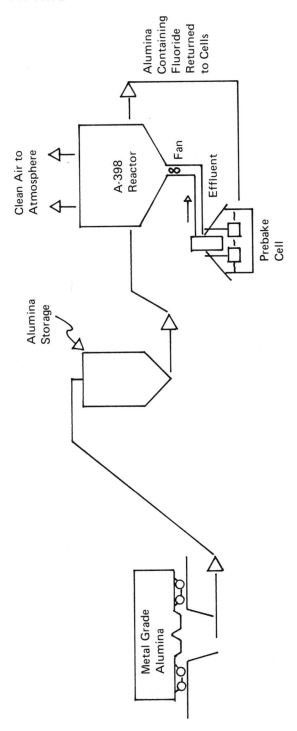

Figure 28. Alcoa-398 Dry Process.

Research continued with alumina contractors, or reactors, designed to handle a larger volume of cell gases and to be installed in the same location as the scrubber equipment or dry process dust collectors. This approach proved to be fruitful (Figure 28).

The Alcoa-398 Process, as it is now known, is composed of a large rectangular reactor containing a fluidized bed of alumina, dust collector, conveyors to convey alumina to and from the reactor for delivery to the smelting cells, associated control apparatus and appropriate alumina storage. The reactor also incorporates a system for feeding alumina uniformly to the various sections of the reactor and an apparatus for discharging alumina from the reactor after it has been fluoridated.

There are no problems associated with handling the wide range of fluoride concentrations from that of Soderberg gases to the highly diluted prebake gases. Using the modified sandy alumina commonly produced within the Alcoa system, extremely high HF efficiency, invariably above 99% and frequently 99.9%, is obtained at alumina feed rate of 50% of cell consumption. The HF in the pot gas is adsorbed by the alumina in the fluidized bed while the particulate fluorides are being intercepted by the alumina. Final interception of particulate, both alumina and fluoride, is a mechanical filtration step accomplished by filter bags located above the reactor. Particulate collection efficiency depends on the filtration ability of the bags. Table 7 shows the

Table 7.   Alcoa-398 Process—Fluoride Collection Efficiencies and Particulate Loss

	F Efficiencies %			Particulate Loss
	Gaseous	Solid	Total	lbs./ton Al
Plant A	99.5	98.3	99.2	0.14
Plant B	99.4	96.3	98.5	1.41
Plant C	99.2	91.8	96.7	4.54

relationship between gaseous and solid fluoride efficiency and the effect of bag filter performance.

With the successful completion of the research and development program, the first production unit was put into service with a new potline at the Badin (North Carolina) Works in 1967. Representative installation and operating costs of major systems in use are shown in Tables 8 and 9.

Work on pot hooding design has continued through the years since the wet or dry systems can capture fluorides only from cell gases which are collected and delivered to the system. In newer potlines, hooding is designed to achieve 95% overall collection efficiency, and performance at less than 95% is considered unacceptable. In normal cell operation, it is necessary to remove a few covers at intervals to replace anodes, tap metal, and perform other routine cell operations. During these periods, there is some fluoride lost to the atmosphere, and the 95% overall efficiency is the net effect of these losses.

Table 8. Typical Costs—Prebake Potline Systems

INVESTMENT COST
($/Annual Ton Aluminum)

	(1967–1970) A-398	(1958–1966) Dry Process	(1951–1956) Precipitators Plus Scrubbers	(1949–1951) Rotoclones Plus Scrubbers
Fans	$ 3.30	$ 3.00	$ 3.10	$ 4.00
Process equipment	23.40	16.50	24.20	13.40
Alumina handling	4.00	5.00	—	—
TOTAL (ACTUAL $)	$30.70	$24.50	$27.30	$17.40
1970 DOLLAR BASIS	33.00	32.75	48.00	36.00

OPERATING COST
($/Annual Ton Aluminum)

	A-398	Dry Process	Precipitators Plus Scrubbers	Rotoclones Plus Scrubbers
Operating supplies	$ .97	$ .50	$ .72	$ .20
R & M labor	1.00	1.25	.35	.30
R & M supplies	.45	.75	.34	.20
Power	1.73	1.00	.72	.60
TOTAL (1970 DOLLARS)	$4.15	$3.50	$2.13	$1.30
F Credit	($8.00)	($7.50)	—	—
Net Cost	$3.85	$4.00	—	—

Table 9. Typical Costs—Vertical-Spike Soderberg Potline Systems

INVESTMENT COST
($/Annual Ton Aluminum)

	(1949–1955) Scrubber	(1952–1955) Wet Electrostatic Plus Scrubber
Fans	$ 2.00	$ 3.00
Process equipment	6.50	17.80
TOTAL (ACTUAL $)	$ 8.50	$20.80
1970 DOLLAR BASIS	15.75	37.20

OPERATING COST
($/Annual Ton Aluminum)

	Scrubber	Wet Electrostatic Plus Scrubber
Operating supplies	$ .31	$ .45
R & M labor	1.40	2.80
R & M supplies	1.00	.70
Power	.19	.54
TOTAL (1970 DOLLARS)	$2.90	$4.35
F Credit	—	—

TODAY'S SITUATION

Since the first A-398 went into service in 1967, four new Alcoa potlines have gone on stream in the United States, all using the A-398. Two more are under construction, and they will use the A-398. In addition, five older potlines have been converted or are being converted from their original process to the A-398. Studies of equipping two of the older plants with A-398 are under way, and, as investment money becomes available or existing facilities fail, A-398 application will be expanded.

In the routine monitoring of our fume control efforts, we measure total fluoride at the inlets to fume treatment facilities and total fluoride at the emission points. Sufficient information, where possible, is also collected to determine unit efficiency for gaseous fluoride and particulate fluorides. Finally, we measure total particulate loss. Where possible, we also sample in the building monitors for fluoride. Using these figures, we are able to obtain both collection and treatment efficiency values and monitor our fume control efforts.

13.
ALCOA-398 PROCESS APPLICATION TO SODERBERG CELLS

C. C. Cook

Technical Manager, Smelting Division

G. R. Swany

Chief Process Engineer, Smelting Pollution Control

J. P. Hupy

Senior Environmental Engineer, Environmental Engineering Division
Aluminum Company of America
Pittsburgh, Pennsylvania

The Alcoa-398 Process is now in use on 14 of 34 prebake potlines in the Alcoa system. The process in the pilot stage, however, was developed on the prototype 110,000-ampere vertical-spike Soderberg cell now in use at the Veracruz Plant of Aluminio, S.A. de C.V., and the Pocos de Caldas Plant of Companhia Mineira de Aluminio. Although the Veracruz Plant originally was equipped with the dry process in which reaction takes place in the filter cake on the bags (the Alcoa-173 Process), the two plants are now fully equipped with the Alcoa-398 Process. In addition, the new potline of Soderberg cells installed by Norsk Alcoa A/S at Lista includes the Alcoa-398 Process for fume treatment. Thus, 4 of the 18 Soderberg potlines operated in the Alcoa system have fume treated by the new process. As in the case with prebakes, the application is to the more modern potlines. Also, as in the prebake application, we have sometimes converted from the Alcoa-173 Process to the Alcoa-398 Process for reasons of efficiency and cost.

HISTORY

The first effort in developing a fluid bed dry process for recovery of fluorides from pot gases was undertaken at Alcoa Research Laboratories at New Kensington, Pennsylvania. The initial efforts were aimed at a pot-mounted unit to allow fume collection, treatment, and alumina heating-drying-fluoridation to be carried out in a chamber designed integrally with the anode-supporting superstructure. The close coupling of the system was expected to reduce costs of ductwork, and, by predrying and heating of alumina, to reduce HF emission from the cell and increase the rate of solution of alumina in the electrolyte. The effort to design a pot-mounted unit quickly revealed that the limited headroom in a potroom was inconsistent with simple equipment geometry. Effort, accordingly, was turned to a floor-mounted fluid bed and baghouse assembly. This prototype Alcoa-398 reactor, 40 inches in diameter, handled a total gas flow rate of 600 SCFM, giving a superficial bed velocity of 1.4 fps at 100° C bed and gas temperature.

In initial performance tests, HF content of gas exiting the system was too low to be detected by the analytical equipment used for the test. Later tests demonstrated efficiencies to be consistently above 99% recovery of combined gaseous and solid fluorides. Fluoride consumption by the pot immediately dropped by about half. Further tests were made using different sources of alumina, varied bed temperatures, varied bed depths, and reduced concentrations of HF in pot fume. The efficiency of the system proved relatively insensitive to all of the variants, and suggested (1) that recovery of HF was by a very rapid reaction,[1] provided that contact of the reactants was adequate; and (2) that the system would recover HF efficiently even from dilute gas streams such as originated in prebake cells.

The 110,000-ampere pot used in these tests was a conventional vertical-spike configuration, with pot gases collected in manifolds and passed through burners which provided for oxidation of most of the combustible species in the gas stream. The burners normally remained hot during the tests, but on the few occasions when the flame went out and remained out for several hours, condensed tars fouled both the bed and the fan. As a result, pilot burners were added to assure that the main burners remained operative. We warned in a previous paper[2] that such burners were mandatory in Soderberg installations, but we have since found that the dilution of the unburned hydrocarbons from a single nonfunctioning burner renders the condensed tars relatively harmless. Only massive failure of the burner system results in tar problems in Alcoa-398 reactors and fans, and an adequate level of attention to burner performance by regular operating personnel is all that is required.

With some encouragement from small-scale tests with dilute fume, we undertook a larger reactor design to handle fume from prebake smelting cells and explore higher bed velocities. Based on this experience, the first commercial installation was made at our Badin, North Carolina, smelter. Since that time, we have expanded Alcoa-398 use in the Alcoa System and have licensed the system for use to another aluminum producer for application to prebake potlines. Only recently, however, have we again turned our attention to application of the process to Soderberg pots.

There are a few areas of major difference in application of the Alcoa-398 Process to Soderberg pots as compared to prebakes. First, since fume volume from VSS Soderberg cells is less, the equipment to handle the gas is much smaller. Both investment and maintenance are inherently less costly. The VSS gas stream bears hydrocarbons in greater concentration than in prebake cells. The Soderberg system does, therefore, require more cleaning. The hydrocarbons do not present any problem in the reactor system. From unpublished work, we know that alumina has a substantial capacity to adsorb hydrocarbons as well as HF gas, and, although we have not applied the Alcoa-398 Process to horizontal-spike Soderberg cells (Alcoa operates none of these), there is reason to believe that a properly designed Alcoa-398 recovery system could be applied to such cells.

INVESTMENT

Because our Soderberg installations of Alcoa-398 fume treatment date only from late 1968, we have limited cost data. Some rather definitive performance tests have been made, though, and we can project performance and cost, we think rather accurately, from the facts available.

First, consider capital cost. Singmaster and Breyer estimates $37.10 per short ton of annual aluminum capacity for installation of the Alcoa-398 system in a new prebake smelter.[3] This figure is in 1970 U.S. dollars, based on a model assuming 2,500 actual cubic feet of air moved per pound of aluminum. (We would emphasize, however, that modern Alcoa designed smelting cells provide in excess of 95% fume collection using only 2,000 ACF air per pound aluminum, and that the Singmaster and Breyer cost figure is, therefore, overstated by about 25% for new Alcoa-designed installations.) In prebake installations of commercial size, the investment cost is nearly proportional to the volume of air moved, and this varies, in the Alcoa System, according to Figure 29. The 2,500 ACF quoted is, in SI units, about 156 m^3/kg aluminum, and the Soderberg air requirment for a comparable size pot is 22 m^3, or about 14% of the prebake pot requirement.

Analyzing the $37.10 figure using internal information, we have assigned the distribution of costs according to Table 10, columns (1) and (2). The individual cost elements can be adjusted according to a scaling formula described beneath the table.

These projections are confirmed by those figures we have available from Latin American installations, but, of course, they will vary with the country and region in which installed. Eftestøl[4] reported costs of 4.9 million n. kr. at Lista, a 50,000 M.T. plant, which would be $12.73 per S.T. Significant factors in cost variation, in addition to location, are (1) reactor size as determined by desired residence time for alumina (this has a significant effect on the reserve capacity of the system in case of alumina flow interruption), and (2) sophistication of the control system, which also bears on reliability of performance. We would expect the distribution of costs among elements of the system to vary only slightly from those listed in the table, regardless of total cost.

PERFORMANCE

Performance data from operating Alcoa-398 systems on Soderberg cells is limited, so far as efficiency is concerned, but consistent. As noted in this

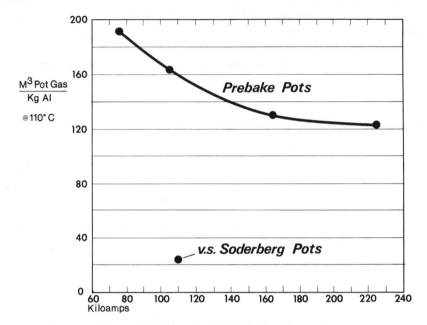

Figure 29. Fume Control Air Requirements

paper before, the prototype reactor recovered more than 99% of all F delivered to it. Efficiencies in monthly tests by Aluminio, S.A. de C.V. at Veracruz, are normally about 99%. Eftestøl *et al.* report[4] 99.9% of fluoride recovered and 98.2% of all dust in tests at Lista over a three-month period in 1971. These reports lead us to conclude that the efficiency of the Alcoa-398 Process as applied to Soderberg cells in commercial operation can be considered to be at least as good as reported for prebake applications.[2] These direct efficiency measurements permit calculation of aluminum fluoride savings as follows (Al Fl means commercial aluminum fluoride containing 90% AlF_3):

The normal concentration of total fluorine in gas emitted from the Soderberg cell as operated in U.S. installations is 800 to 1,000 Mg/Nm^3, with an average around 900. Gas collection systems normally handle about 22 m^3 gas per Kg of aluminum. Assuming 99% of the fluorine recovered, aluminum fluoride savings would be:

$$\frac{Kg\ Al\ Fl}{Kg\ Al} = \frac{900\ MgF}{Nm^3\ gas}\frac{298\ Nm^3}{383\ m^3}\frac{22m^3\ gas}{Kg\ Al}\frac{Kg}{10^6mg}\frac{.99}{.9}\frac{84\ \#\ AlF^3}{57\ \#\ F} = 0.025$$

Table 10. Alcoa-398 System Capital Costs

	(1)	(2)	(3)	(4)	(5)	(6)
	Prebake Pots		Cost Factor		Soderberg Pots	
	%	$/ST Al	x	(1/7x)	%	$/ST Al
Reactor package	56.5	$20.96	0.8	0.21	42.0	$ 4.40
Alumina handling	8.8	3.26	0.6	0.31	9.6	1.01
Alumina storage	4.0	1.48	0.1	0.82	11.5	1.21
Elec. and Control	13.6	5.05	0.8	0.21	10.1	1.06
Paint and misc.	2.6	.96	0.2	0.67	6.1	.64
Fdns. and paving	6.0	2.23	0.9	0.17	3.6	.38
Engineering	3.9	1.45	0	1.00	13.8	1.45
Direct cost	95.4	$35.39			96.7	$10.15
Site prep.	0.5	.19	0.9	0.17	0.3	.03
Power subst.	0.8	.30	0.9	0.17	0.5	.05
Const. overhead	3.3	1.22	0.8	0.21	2.5	.26
	100.0	$37.10			100.0	$10.49

Cost reported—Veracruz and Pocos de Caldas   9.43

Cost reported—Lista   12.73

Explanation of Cost Factor
Columns (3) & (4)

$$C_{PB} = C_S R^x$$

where C_{PB} is the cost of an installation R times larger than a comparable installation which costs C_S; and ^{x} is an exponent with a value of 0 to 1. (A value of x equal to zero would reflect no scale advantage at all. For example, we consider the cost of engineering essentially the same, regardless of equipment size, so with x = 0, $C_{PB} = C_S$ for engineering costs.) Also, note that the common use of the scaling formula is to scale up costs. Since we are scaling size and costs down, we restate the formula:

$$C_S = C_{PB} \frac{1}{R^x}$$

and the cost scaling factor from prebrake to Soderberg fume control equipment becomes:

$$\frac{1}{R^x}$$

The power factor x is selected in each case based on experience and the projected figures for installation of an Alcoa-398 system to treat Soderberg pot fume are tabulated in columns (5) and (6).

Table 11. Alcoa 398 System Operating Costs

	(1)	(2)	(3)	(4)
	Prebake Pots		Cost Factor	Soderberg Pots
	S & B	Alcoa		
Operating labor and materials	$ 0.34	$ 1.10	.50[a]	.55
Electric power (6 mills/ KWH)	1.67	1.67	0.14^b	.24
Maintenance labor and materials	1.86	1.10	$.25^c$	.28
Royalty	0.64	.64	1.00	0.64
Administration (5% cap. cost)	1.86	1.86	0.28^d	0.52
Taxes and insurance (2% cap. cost)	0.74	.74	0.28^d	0.21
Depreciation (8% cap. cost)	2.97	2.97	0.28^d	0.83
Interest (8% cap. cost)	2.97	2.97	0.28^d	0.83
	$13.05	$13.05		4.10
Credit: Alumina values	(1.07)			(.10[e])
Fluoride values	(10.14)			(10.00[f])
	(1.84)			(6.00)

a. Based on reported experience, Veracruz.
b. Ratio of volume of pot fume treated.
c. Based on reported experience, Veracruz.
d. Ratio of investment from Table 1
e. From Singmaster & Breyer.
f. Based on 50 lbs. Al Fl/ton @ 20 cents/lb

This figure 0.025 Kg Al Fl/Kg Al can also be stated as 50 lbs. Al Fl/ton Al. In a previous paper,[2] we attributed savings of 30 lbs. fluorine per ton of aluminum to application of the Alcoa-398 Process to prebake smelters. This figure equates to 49 lbs. Al Fl/ton Al. Thus, we check the comparability of the Alcoa-398 Process between prebake and Soderberg pots. As a final check, a careful materials balance between two potlines was made over a 39-day period.[5] One line was equipped with an older and less efficient dry process. Commercial aluminum fluoride consumption was less in the Alcoa-398 line by 0.015 Kg/Kg Al or 30 lbs/ton Al.

MAINTENANCE AND OPERATING COSTS

Maintenance of the Alcoa-398 system falls into two general categories: the motive equipment for alumina and pot fume, and fume treatment equipment proper. Maintenance of the former, including conveyors, elevators, and fans, is the province of the maintenance department of most Alcoa smelters. Maintenance of the reactors, ductwork, and baghouses is primarily clean-up work and bag changing, and is considered "operating" labor. There is, in fact, no operating labor as such in the process, since the reactors are fully automated and instrumented and can be monitored remotely by potroom supervisors or computer.

Operating costs for the Alcoa-398 Process were estimated by Singmaster and Breyer[3] at $13.05 per ton Al for a prebake smelter with costs distributed as in column (1), Table 11. We endorse this estimate except for the distribution between operating and maintenance costs, which is largely a matter of accounting. Alcoa's adjusted estimate appears in column (2), Table 11. Appropriate factors have been used to estimate Soderberg operating costs, as explained in the table. One comment regarding the factors used to adjust operating and maintenance costs: the 50% factor used for operating materials and labor is not equally applied between the two. The operating labor cost for the Soderberg application is about 75% of that of an equivalent prebake operation. Operating materials (mostly filter bags) are only 25%. On the other hand, in the case of maintenance labor and materials, the 25% factor approximates both the ratio for labor and for materials between the two systems. Labor may be slightly more (25 to 30%) and materials slightly less (20 to 25%). In no case is cost of cleaning ductwork inside the potrooms included.

METAL PURITY

One last comment on the Soderberg application of Alcoa-398 Process regarding metal purity. We reported earlier that there was no measurable loss of metal purity associated with the process.[2] Eftestøl reports pickup of Fe_2O_3 in alumina processed in Alcoa-398 reactors at Lista as 0.023.[4] Careful scrutiny of numerous grade sheets at Veracruz reveals Fe pickup in metal in pots employing Alcoa-398 treated aluminas as about .01 to .02. No evidence of silicon pick-up was observed.[5]

CONCLUSIONS

In summary, we conclude that the Alcoa-398 Process is applicable to prebake cells or Soderbergs of vertical-spike design. We believe the process could be adapted to horizontal-spike Soderberg cells. Pilot burners on VSS cells are no longer considered necessary, though some extra precautions are required to keep burners functioning. Capital and operating costs for VSS systems are about 30% of those for a prebake smelter of comparable tonnage. Metal purity losses due to the process are confined to about .01% as Fe pickup in metal. In either prebake or Soderberg plants, premium metal purity can be maintained in selected pots by use of less than 100% feed of alumina to reactors as long as adequately active alumina is supplied.[1]

EPILOGUE

Alcoa has continued to study alternate contacting systems for chemisorption of HF in the pot gas stream on incoming alumina. One objective, of course, has been to reduce interior surfaces upon which tars and soot build up when Soderberg pot gases are carried. We have successfully operated a dense phase injection system, achieving in excess of 99% HF recovery and at a somewhat reduced pressure drop from that in the Alcoa-398 Process system. Studies of this new system will continue. If the new method proves superior in operating costs and convenience, we will be prepared to extend it to additional Soderberg plants.

REFERENCES

1. Cochran, C.N., W. C. Sleppy, and W. B. Frank. "Fumes in Aluminum Smelting; Chemistry of Evolution and Recovery," Journal of Metals (September, 1970), 3—6.
2. Cook. C. C., G. R. Swany, and J. W. Colpitts. "Operating Experience with the ALCOA-398 Process for Fluoride Recovery," Journal of the Air Pollution Control Association (August 1971), 479—483.
3. Independent Study by Singmaster & Breyer (235 East 42nd Street, New York, N.Y. 10017), under sponsorship of U.S. Environmental Protection Agency, Office of Air Programs.
4. Eftestøl, T., Ej Mørkesdal, and A. Kj Syrdal. "Duplex Gas Cleaning System at a Modern V.S. Soderberg Plant." Paper presented to AIME annual meeting, San Francisco, California, February 20—24, 1972.
5. Private communications with N. Clay Cook, Aluminio, S.A. de C.V., Veracruz, Mexico.

14.
AIR POLLUTION CONTROL
ELECTRIC ARC MELTING FURNACES

Frank R. Culhane

Vice-President
APC-Products
and

Clyde M. Conley

Metallurgical Consultant
Air Pollution Control Division
Wheelabrator-Frye Inc.
Mishawaka, Indiana

Today's state of the art with regard to direct arc electric melting fume collection covers over 30 years of in-plant fume control and 20 years of effective air pollution control experience. The spade work has been done, and today industry is displaying effective control results with justifiable pride for in this country, alone, over 250 electric arc furnaces in the foundry and steel industry have been equipped with effective control equipment. The results are well documented, and this paper will review the state of the art with regard to electric furnace fume control and identify whatever trends are evident.

ELECTRIC FURNACE FUME

What are we attempting to control? We use the term electric furnace fume, but what is it? How is it generated? And what is its nature? By sampling and examination we know that it consists of solid particulate matter. We know, certainly, that it is formed at extremely high temperatures within the furnace. The metallurgists tell us[1] that at these high temperatures iron will vaporize, and, as the iron vapor rises above the bath, it oxidizes, cools, and solidifies into minute submicron particles as small as 0.01 micron in diameter, which we define as fume. In addition to fume-size particles, much larger particles of dust, dirt, and rust are carried out of the furnace into the atmosphere by the gases generated within the furnaces.

The chemistry of electric melting furnace fume is predominately iron oxide but varies during the different phases of the heat depending, of course, upon the scrap charged and metal being produced. Zinc from any galvanized metal that may be in the scrap will fume at relatively low temperatures, and in these furnace conditions the fume will contain a high percentage of zinc oxide genererated during the initial phase of melting. During silicon reduction, white silica fume is generated. As silicon monoxide is more volatile than either silicon or silicon dioxide, it has been postulated that silicon monoxide evolves as a gas, and, after striking the air, it immediately oxidizes, cools, and condenses into spherical submicron particles of silica fume. Silica fume particles generated in submerged arc furnaces range from 0.01 to 0.5 micron, with a mean particle size of 0.06 micron. And we would anticipate gas-borne silica fume from direct arc furnaces to be in the same size range.

Figure 30. Electron micrographs of Electric Furnace Fume[2]

PARTICLE SIZE

Let us look at electron microscope photographs[2] of electric furnace fume as shown in Figure 30. Only one or two particles are as fine as 0.01 micron.

Gas-borne solids smaller than 0.01 micron are difficult to find in any large quantity due to agglomeration resulting from Brownian movement of the particles. A four-particle agglomerate can be seen in the photograph which acts as a one micron particle. The largest particle is four microns in one plane and it appears to be at least a four-particle agglomerate, the largest of which has jagged edges like a flake of rust. From the sample in the photograph it can be seen that 100% of these particles are below 5 microns, 95% are below 2 microns, and the mean size is 0.5 micron. It is this extremely fine particulate matter that must be separated from the furnace off-gases to prevent pollution of our working atmosphere.

QUANTITY

The quantity of particulate matter has been checked from numerous control systems, and the amount varies from a low of 5 pounds of fume per ton of charge to a high of 55 pounds per ton, with an average of 20 pounds per ton of charge, depending upon scrap conditions, oxygen lancing practices, and the carbon content of the bath at the beginning of lancing. The density immediately after recovery in a dry state and before deaeration, which occurs with time, averages 35 pounds per cubic foot.

CHEMISTRY

The chemical composition of the fume generation varies widely from heat to heat and from phase to phase of a single heat. Table 12 shows a chemical analysis between the different phases of a single heat. Notice that for a single heat, the iron oxide varies from 26.6 to 56.75%, and silicon dioxide varies from a trace during reduction to 9.77% during melting. Scrap, temperature, slab conditions, and the furnace additions are contributing factors to the chemical variance of electric furnace fume.

FURNACE OFF-GAS

The makeup of the off-gas consists of infiltrated air plus combustible gases created by (1) electrode consumption, (2) vaporization and disassociation of oil and water in the scrap, (3) melt-down of the scrap, and (4) decarbonization.

During melting, a typical gas analysis[3] will run 8% carbon monoxide, 4% oxygen, and the rest nitrogen. After the second and third charges, peak values of carbon monoxide in the range of 71.5% and hydrogen values up to 38% have been reported. During oxygen lancing, the off-gas will run approximately 30% carbon monoxide, ½% oxygen, and the rest nitrogen. Peak carbon monoxide values during lancing at 1% carbon bath have been reported as high as 87.95%.

Table 12. Chemical Analysis of Electric Furnace Fume during Various Phases of the Heat[3]

Dust Composition

Phase	% SiO₂	% CaO	% MgO	%[a] Fe₂O₃	% Al₂O₃	% MnO	% Cr₂O₃	% SO₃	% P₂O₅
Melting	9.77	3.39	0.45	65.75	0.31	10.15	1.32	2.08	0.60
Oxidizing	0.76	6.30	0.67	66.00	0.17	5.81	1.32	6.00	0.59
Oxygen lancing	2.42	3.10	1.83	65.37	0.14	9.17	0.86	1.84	0.76
Reduction	Tr.	35.22	2.72	26.60	0.45	0.70	0.53	7.55	0.55

a. The iron content was determined as total iron and converted to Fe_2O_3.

The high hydrogen content of the gases during the early phases of melting the second and third charges and the high carbon monoxide content during oxygen lancing clearly present the possibility of an explosive gas mixture, and the control system *must* be designed to minimize this hazard.

For the most part, self-ignition of the gases plus an adequate indraft of secondary air has been used in this country to render the gases safe. By assuming 100% conversion of oxygen to carbon monoxide during oxygen lancing and assuming 85% combustion of the carbon monoxide to carbon dioxide in the system plus 100% excess air above theoretical, safe working conditions are assured for cold charge furnaces. Care must be exercised in designing the exhaust system, however, as carbon monoxide peaks in excess of 100% theoretical conversion of oxygen to carbon monoxide have been measured during lancing.

GAS QUANTITY

The quantity of gas generated and the temperature of the gas vary, and vary rapidly. Care must be exercised in establishing the proper exhaust rate based on careful consideration of the following factors:

1. the charge makeup;
2. the rate of combustion of the combustible materials in the charge;
3. rate of converting solid scrap to molten metal;
4. anticipated transformer usage;
5. iron ore additions;
6. lime addition or limestone additions;
7. hot metal addition;
8. carbon content of bath at beginning of lancing;
9. oxygen lance rate;
10. alloying additions;
11. carbon additions;
12. rate of decarbonization; and
13. method of fume capture.

Correlation of these factors and the quantity and temperature of fume generation are normally determined by the fume control equipment manufacturer or by experienced engineers. Presentation of the relationships between these factors and the gas temperature and quantity if included in this paper would necessarily require oversimplification which, should it be misapplied, could be disastrous.

CLEANING EQUIPMENT

To date, the tools of particulate air pollution control for electric furnaces have been wet scrubbers, the electrostatic precipitator, and the baghouse. Let us review each one separately.

Scrubbers

Considerable experience has been acquired throughout industry in the application of many designs of wet dust collectors to collect many different kinds of dust from a variety of sources. From time to time, most of these have been applied to electric furnace fume with limited success. Common to all these designs is the relative amount of energy (2 to 4 hp per 1000 acfm) expended to intermix the gas and the water, resulting in collection efficiencies from 25 to 75%, with the higher efficiencies proportionate to the expended energy. The Venturi scrubber, which was developed in the mid 40s, increased the energy input to higher levels than ever before (10 to 14 hp per 1000 acfm), with a corresponding increase in efficiency.

The operating principle of the typical high-energy scrubber relies on the introduction of low-pressure water into a zone of high gas velocity in the range of 12,000 to 24,000 fpm. This high gas velocity atomizes the water, forming a tremendous number of slow-moving droplets, which creates a tortuous passage for the fast-moving fume particles and exposes a large surface area of water, resulting in optimum conditions for particle impingement into the water.

The higher the gas velocity, the higher the cleaning efficiency (Figure 31 shows the relationship between pressure drop and residual particulate loading for a Venturi scrubber). For a residual loading of 0.05 grain per scfd, 40 in. WG is required. For 0.02 grain per scfd, 60 in. WG is required. This curve applies to oxygen-blown converters. A correction factor for electric furnaces probably is required, but the relationship is undoubtedly close.

It is estimated that, in the foundry and steel industry, approximately 20 furnaces have been equipped with high-energy orifice or Venturi-type scrubbers. Regardless of their apparent simplicity and advantages, they are not the most predominant cleaning device used on arc melting furnaces.

The anticipated advantages of the Venturi are (1) solids are collected in a wet state, (2) the Venturi is less susceptible to temperature overruns, and (3) simplicity.

Recognized disadvantages are (1) higher operation cost in the range of 300% over the baghouse; (2) potential water pollution problems if the sludge handling system should malfunction, (3) potential imbalance and noise problem due to high tip speed of fan impeller, and (4) potential corrosion problems due to sulphur and oxygen.

The lower operating temperature of the high-energy scrubber reduces the volume of the gas to be cleaned, which reduces the size of the ductwork and scrubber. This reduction is offset by increased cost of the high-pressure fan, larger motor and starter, and water-handling equipment. The best comparison indicates that the high-energy scrubber including the sludge-handling equipment is approximately 15% more expensive than a baghouse on a large furnace. Without the sludge disposal system, the price of the Venturi scrubber is

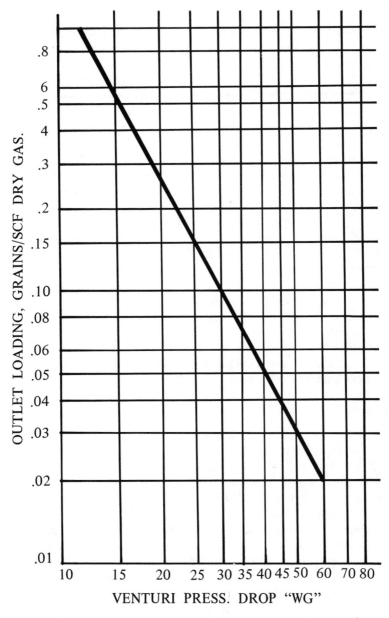

Figure 31. Cleaning Efficiency *vs* Pressure Drop, Venturi System[4]

approximately 5% less. On smaller furnaces, the high-energy scrubber appears to be appreciably more than the baghouse.

Electrostatic Precipitators

In view of the wide acceptance of the electrostatic precipitator as a means of effective gas cleaning throughout industry, it is surprising to find few in use as pollution prevention devices in electric furnace shops today. In this country, none are in service in the foundry industry and only one is in service on electric furnaces in the steel industry. The well-documented reason is graphically demonstrated in Figure 32, which shows the electrical resistivity of iron oxide from various oxygen lance steel-making processes. Particles that have a high resistivity are difficult to precipitate electrically, and difficulties are often encountered above 10^{11} ohm-cm. As seen in the figure, electric furnace fume is highly resistive and varies from 10^{11} at 600° F to 10^{13} at 300° F. This value can be improved by the introduction of water into the gas stream, and evaporative conditioning towers ahead of the precipitator are used to stabilize the gas temperature and humidity. To do this, a sufficient source of steady heat should be available to properly control gas humidity. The off-gas temperature from an arc furnace rises and falls sharply during a given heat, and desired humidification is readily obtainable during high temperature peaks but not during periods of low temperature. Steam and/or highly atomized water introduction at times of low temperature would result in condensation difficulties in the system. The lower resistivity curve for the open hearth furnace fume is attributed to the water vapor resulting from combustion of the furnace fuel as well as the sulphur trioxide in the gas.

For the above reasons, the gas cleaning industry generally does not regard the application of electric precipitators for cleaning arc furnace gases as ideal.

Baghouses

Product recovery from stack gases by the use of filter fabrics has resulted in the development of an effective means of gas cleaning often referred to as a baghouse. This method of gas cleaning represents one of the best means of obtaining the highest consistent recovery of gas-borne particulate matter. Production requirements have demanded reliability and durability of design, which has been achieved.

Surprisingly enough, a baghouse, or a fabric collector if you prefer, represents a rather versatile device for the abatement of particulate air pollution. It has a uniform high collection efficiency regardless of varying temperature, varying moisture, varying dust load, and varying volume. Field efficiency tests handling submicron electric furnace fume in the range of 1 grain per cubic foot fume load have consistently resulted in efficiencies in excess of 99% by weight and an effluent free of visible solids.

Today, there are about 200 baghouses in operation in the foundry and steel industry handling electric furnace fume. Most of these units in the steel industry incorporate the features shown in Figure 33, consisting of tubular bags open at the bottom and closed at the top with the gas entering the

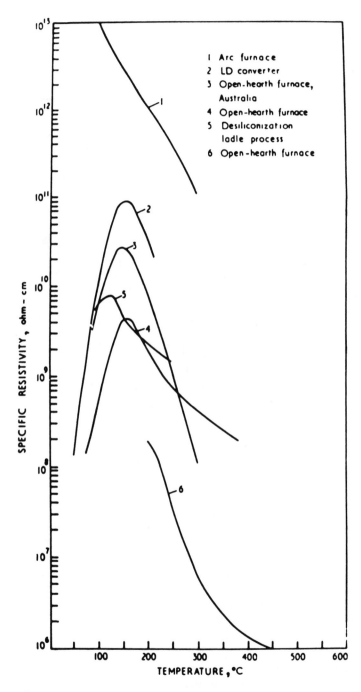

Figure 32. Electrical Resistivity of Red Oxide Fume from Various Oxygen-blown Steelmaking Processes[5]

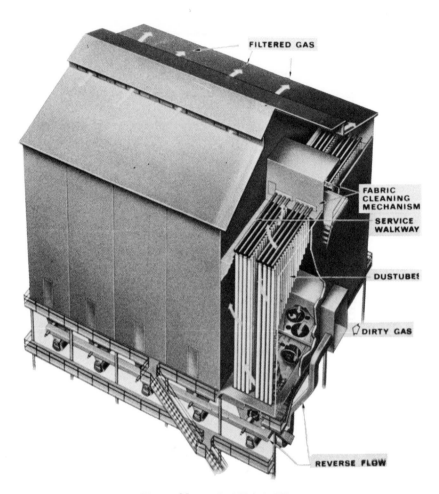

Figure 33. Typical Fabric Filter

bottom of the unit and flowing upward into the tubes and through the fabric, depositing the fume inside the bags. The fume is removed from the bag by interrupting filtration, then reversing the flow or shaking.

In multifurnace shops operating around the clock, a maintenance compartment is normally provided to allow maintaining the unit while in service without adversely affecting filtration efficiency. Filter bags can be inspected and maintained by average maintenance personnel. The bags are accessible from internal compartment walkways, and the bag grouping allows for ease of bag inspection and changing. The compartment damper valve design provides positive and complete interruption of gas flow without leakage during periods of cloth cleaning and compartment inspection. A suction unit will prevent

Figure 34. Fabric Filter on Steel Foundry

erosion of the fan scroll and impeller and will provide a cleaning and more desirable unit, but at the disadvantage of increased capital cost. During inspection of a compartment in a suction unit, a partially open inlet damper valve will allow for infiltration of outside air into the compartment to positively purge furnace gas from the compartment.

The advantages of a fabric collector are (1) efficiency—greater than 99% by weight, producing an effluent free of visible solids; (2) maintainability during operation; and (3) rugged simplicity.

The disadvantages are (1) space; (2) bag life; and (3) requirement of over-temperature protection equipment.

A typical steel foundry installation can be seen in Figure 34, which shows a 14-gauge sheet metal pressure baghouse equipped with 5 in. diameter x 14 ft. long bags handling four electric arc furnaces. The separate stacks in each compartment are free of visible discharge. This unit is equipped with acrylic bags, and side-draft hoods are utilized for furnace fume pick-up.

A typical steel mill installation can be seen in Figure 35, which shows a structural steel baghouse equipped with 8 in. diameter bags 22 ft. long. Ductwork in the foreground leads from the canopy hoods over the furnaces and tapping aisle to four fans which discharge into the baghouse in the background. The system ventilates furnaces in a furnace shop. The canopy hood as shown in Figure 36, used in combination with the baghouse, has met with full approval of the local neighbors and acceptance by the authorities and has demonstrated the ability of the baghouse to comply with local air pollution control codes.

Figure 35. Fabric Filter on Steel Mill

Figure 36. Canopy Hood Duct and Fabric Filter on Steel Mill

Cloth filtration of furnace gases requires gas cooling to protect the filter fabric and insure economical fabric life. After adding primary and secondary combustion air to the furnace off-gas, the bases at flame temperature are cooled to 550° F for Fiberglas or 275° F for either Orlon or Dacron.

If a close-fitting furnace roof hood is used for furnace fume control, the amount of infiltration air, resulting from sufficient indraft to capture the fume, also tempers the furnace gases to within the safe operating temperature range of Orlon or Dacron. This same principle applies to the canopy hood.

Gas cooling for direct steel evacuation can be achieved by tempering air and/or radiant cooling, and/or spray cooling. On ultrahigh-power furnaces, two basic and wholly different systems are evolving, depending on spar usage.

On furnaces using 5 pounds and over of spar per ton charged, Dacron is being used, and the preferred cooling system consists of a convection cooling by water jacketed ductwork from flame temperature to 1200° F, radiant cooling from 1200 to 375° F, followed by tempering air from 375 to 275° F. The cooling is completely dry, and the mass of the system provides a heat sink which minimizes temperature surges at the fan and collector. The radiant cooling plus tempering air is designed for summer conditions. During the winter months, the tempering air is modulated to prevent overcooling.

On low spar usage, under 5 pounds, Fiberglas is used, which reduces the gas cooling requirements appreciably, a preferred cooling trend consists of spray cooling following the furnace elbow. The gases are cooled from flame temperature to 600° F by evaporative cooling, followed by, in the case of multifurnace shops, mixing to 550° F. With the trend to increased oily scrap and ultrahigh power, this system is attractive as it has a faster rate of cooling response and greater thermal control capacity during periods of excessive temperature surges. If the spray tower is used, the baghouse should be fully insulated.

There are many interrelated costs in the various combinations of gas cooling. The trend toward higher and higher baghouse operating temperatures reduces cooling requirements but increases the cost of the gas cleaning equipment. Therefore, regardless of the higher operating limits of new fabrics, considerations should be given to a balance between the capital cost of gas cooling and gas cleaning. If cooling is carried to an extreme to reduce volume, overcooling can occur, resulting in operating difficulties as the dew is approached.

DISPOSAL

A problem that has not been satisfactorily solved nor adequately considered is fume disposal. A 300,000 annual ingot ton shop generates an average of 3,000 tons of fume per year. As more and more shops install effective means of gas cleaning, a sufficient increase in the supply of furnace fume will warrant special handling and treatment. With today's interest in prereduced pellets, economics may soon permit beneficiation and conversion to metalized pellets for recharging. Certainly, disposal clearly needs and will shortly demand high level research and development in order to provide a better answer than what is available today.

SUMMARY

Electric furnace fume control is neither simple nor insurmountable, but is a highly technical problem requiring engineered answers. The fundamental spadework of the 1950s, industry accomplishments of the 1960s are being displayed with justifiable pride. However, much remains to be done and can be done effectively and economically, provided the control equipment is properly designed and sized with adequate allowance today for tomorrow's increase in furnace productivity.

REFERENCES

1. Turkdogan, E. T., et al. "The Formation of Iron Oxide Fume," Journal of Metals (July 1962), 521–526.
2. Allen, G. L., et al. "Control of Metallurgical and Mineral Dusts and Fume in Los Angeles County, California," Bureau of Mines Circular 7627, April 1952.
3. Harms, F., and W. Riemann. "Measurements of the Amounts of Off-Gas and Dust in 70-Ton Electric Arc Furnaces with Partial Use of Oxygen," Stahl and Eisen (September 27, 1962), 1345–1348.
4. Willet, H. P., and D. E. Pike. "The Venturi Scrubber for Cleaning Oxygen Steel Process Gases," Iron and Steel Engineer (Year Book, 1961), 564–569.
5. Watkins, E. R., and K. Darby. "The Application of Electrostatic Precipitation to the Control of Fume in the Steel Industry," The Iron and Steel Institute, Special Report 83, p. 27.

15.
THE IRON FOUNDRY INDUSTRY

P. S. Cowen

Assistant Technical Director
Gray and Ductile Iron Founders Society, Inc.
Cleveland, Ohio

Emissions and their control in the iron foundry industry are a complex situation that appears clouded by considerable confusion. The chief documented source of emissions is the iron melting process, of which there are several. This is a process common to all casting operations. Secondary and significantly minor sources, by comparison, are either not common to all foundry operations or have had a history of emission control antedating external concern over such problems.

The cupola, a principal melting furnace of the industry, is a matter of interest because of its emission peculiarities. Electric direct arc furnaces, when used as melters and not duplexers, are a source of particulate emissions that require control considerations or techniques. Indirect arc electric, induction electric, and reverberatory furnaces are not sources of interest in terms of emission controls.

An auxiliary iron foundry operation, such as centralized sand conditioning and shakeout, can be an emission source that is of concern, depending on operating technique. The older, so-called windrow and heavy floor shop methods do not have the potential of sand conditioning and shakeout emissions of their mechanized counterparts.

Coremaking methods, materials, and techniques are becoming increasingly more varied. The newer approaches are largely an outgrowth of chemical process technology. Some of both the older and newer core binder technologies can be significant emission problems, others are totally devoid of emission problems, and, naturally, some are within both extremes. These sources require separate discussion as to their reason for use and degree of interest, and are not discussed further.

Cupola emissions that can be of interest, academically and otherwise, are particulates, sulfur oxides, carbon monoxide, and opacity. The first and last are the most troublesome for various reasons. Sulfur compounds and carbon monoxide are ordinarily of no particular concern. Control systems utilization is a function of both rule requirements, effluent of interest, and operating parameters. Following is a description of effluent characteristics which includes control considerations.

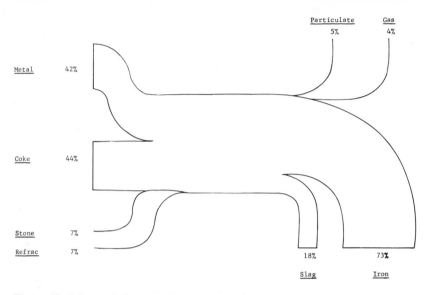

Figure 37. Schematic Process Flow Diagram for One Ton of Iron through Cupola

PARTICULATE MATTER

Particulate matter emissions are generally indicated as being from 17.4[1] to about 20[2] pounds per ton of iron melted from cupolas. A process flow diagram including particulate emissions is shown in Figure 37. The value varies considerably and is mostly dependent on ratios of iron melted to coke, which is the metallurgical fuel. An extensive study by Gerhard Engels[3] produced a regression analysis (see Figure 38) showing the coke input factor as being significant.

Characteristics of the cupola particulates are somewhat unusual compared to most natural and man-made sources. The chemical constituents are indicated in Table 13. As can be expected from metallurgical discipline, the principal matter is silica, iron oxide, and carbonaceous combustible matter. Principal sources of this matter, respectively, are sand adhering to recycled metal, rust, and the friable portion of the coke. Extraneous matter, such as machine shop cutting fluids on steel turnings, nonferrous contaminants, oil bearing scrap, and coated metals, occasionally appears in the form of particulate constituents.

The size distribution of cupola particulate matter is wide ranging and caused by the relatively high flow rates of the carrier gas. An example of a distribution is shown[5] in Figure 39. Obviously, the coarser sized matter either falls out of the effluent or is easily collected. The portion that is 10 microns and smaller has a density and shape comparable to irregular silica particles and is the suspendable matter of interest in ambient air considerations. This portion lies in a normal range of 5 to 20% of the particle size distribution.

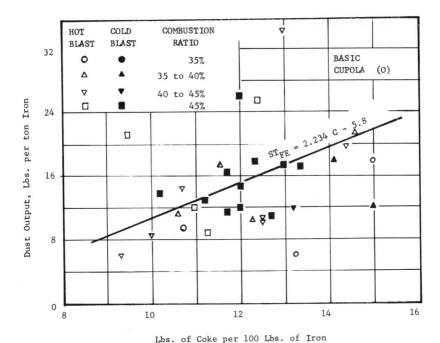

Figure 38. Relationship between Coke Rate and Total Dust Output per Ton of Iron

Table 13. Chemical Composition of Cupola Dust[4]

	Mean Range %	Scatter Values %
SiO_2	20–40	10–45
CaO	3–6	2–18
Al_2O_3	2–4	0.5–25
MgO	1–3	0.5–5
FeO (Fe_2O_3, Fe)	12–16	5–26
MnO	1–2	0.5–9
Ignition loss (C, S, CO_2)	20–50	10–64

Control techniques for particulates from cupolas require the consideration of some operating conditions and integral factors imposed by other effluents and matter. Chief among those under consideration is the engineering design for control of the carrier gas temperature. Depending on the process operation and major control interest, the designer must ideally make choices from narrowed alternatives.

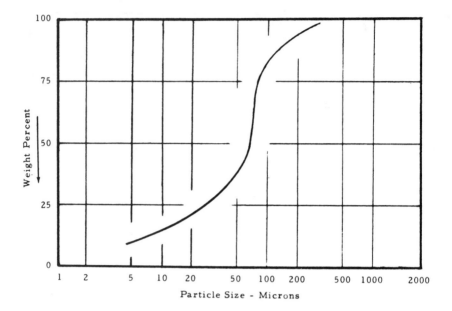

Figure 39. Particle Size Distribution by Weight Percent

SULFUR COMPOUNDS

Sulfur oxide emissions from cupolas have been largely ignored for several reasons, mainly because they are largely undetectable. An explanation as to why this condition exists is necessary to avoid consternation. There can be situations where the presence of sulfur compounds in even minute quantity can influence certain design criteria for emission control. In such situations, the interest is not necessarily in abating that type of emission.

Sulfur is present in iron melting principally in the ingoing coke and metal.[6] There is also a small amount that is naturally present in minor process input. Foundry coke is a rather high-grade, expensive metallurgical fuel that contains up to 0.60% sulfur. Compared to other forms of fuels, it contains a very small amount of this element. Charge metal can typically contain from 0.12% to almost zero sulfur.

Next to water, molten iron is the second most potent solvent. It can dissolve most of the common elements on contact, and, like water, it is not volatile. This particular characteristic of iron can be a nuisance to iron melters when they wish to produce low-sulfur iron. Certain methods of cupola operation to produce low-sulfur iron influence choices of methods of emission abatement. An example is the inability to use Fiberglas filtration on a basic cupola using flourspar as a slag thinner.

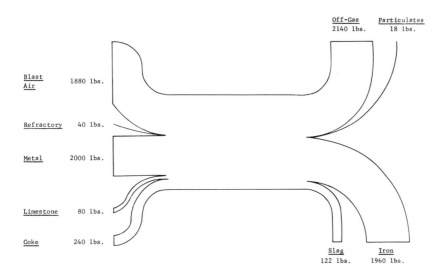

Figure 40. Schematic Flow Diagram for % S Entering (left) and Existing
        Cupola

For the sake of providing a view of the sulfur situation in cupola melting, a process balance for this element is shown in Figure 40. The flow diagram is representative of most operations. Essentially, 91% of the ingoing sulfur exits the process in either the molten iron or slag. Attempts to analyze exhaust gases for sulfur oxides are usually fruitless unless geared toward concentration ranges of 20 to 100 ppm. Therefore, it is usually necessary to arrive at a value by process difference.

CARBON MONOXIDE

Carbon monoxide is an essential part of the melting atmosphere of a cupola. It is interesting to observe that this gas has been ignored in favor of great public concern over particulate matter and opacity. Carbon monoxide has been of interest to foundrymen from the standpoint of industrial hygiene and its known lethal effects under faulty operating conditions. It is probable that the odorless and colorless characteristics of carbon monoxide contribute to its lack of glamour or publicity.

There is confusion in evaluating the status of carbon monoxide in cupolas. A profile of a representative metallurgical flow is show in Figure 41. It must be noted that this profile is for the below-charge opening reaction conditions and not for emission flow. The potential presence and conditions for control are partially or wholly governed by the below-charge opening characteristics and the charge opening.

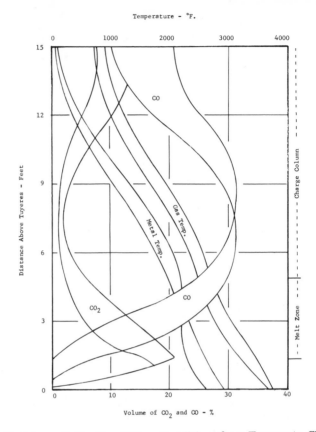

Figure 41. Schematic Profile of Flow Conditions from Tuyeres to Charge
Opening

Table 14. Ignitability of Cupola Top Gases with a Moisture Content of About 8%[7]

Gas Temperature °C.	Combustion Chamber °C.	With/Without Igniting Flame	Minimum CO Content %
20	420	with	13
20	500	with	12
20	600	with	10
210	400	without	15
210	500	without	13
210	600	without	12
200	400	with	8
200	500	with	7
200	600	with	6

Table 15. Orsat Analysis of Cupola Exhaust Gases[8]

	Cupola A	Cupola B	Cupola C
Light Off Period:			
CO_2	0.3–0.8	1.0–10.3	0.8–3.0
O_2	19.3–20.3	10.1–19.2	17.4–20.5
CO	0.0	0.0	0.0
Melting Period:			
CO_2	2.0–5.4	10.0–13.4	2.2–5.1
O_2	14.4–18.5	6.5–10.1	14.5–19.2
CO	0.0	0.0	0.0

Combustion limits for carbon monoxide in air have long been known. The lower limit for self-ignition[5] is 12.5% at 1128° F. Recent research[7] on cupola effluent has produced the limits of ignitability (see Table 14). This helps explain the self-igniting characteristics of most cupola stacks[8] as shown by test results for the gas (Table 15).

It should be noted that inspirated air through charge openings is essential to the control of carbon monoxide. In original designs of standard cupolas, the inspirated air was controlled by relative sizing of stack height and charge openings. This provided protection from poisoning hazards. The advent of mechanized charging and emission controls disrupted this old-fashioned balancing.

The use of afterburners as a method of carbon monoxide control is actually limited to one of being an ignition source. There are obviously conditions where such burners are not necessary. For certain types of secondary combustion control, the igniters are necessary. This includes use as a source of artificially raising exhaust stack temperatures for downstream gas humidity control.

Maintenance of secondary combustion has an effect on particulate matter concentration. The combustibles portion of the particulates originates from the friable characteristics of the coke. In certain operations, there can be a contribution of combustible matter from charge material surface adherents. Sufficient time at temperature can be used to burn these combustibles. This provides a particulate matter reduction that can range from 10 to 64%. Certain jurisdictions use this method as their criteria for emission control in lieu of filtration approaches.

OPACITY

Opacity does not become a common problem on cupola emissions until control equipment is placed in operation. The particulate size distributions and concentrations rarely produce visibility reductions. Opacity is not quantitatively related to emissions[9] and can be considered only as a quality of its

Figure 42. View of fabric filtration filter downstream from cupola gas treating system. Exhaust for effluent is broadside through opening at upper right of structure.

own nature. Conditions for quantitative relationship for opacity to quantity are heavily weighted with restrictions.[10] The effects of condensing moisture in an effluent plume have an obvious bearing on opacity. Depending on social conditions, this form of emission is often of concern and is prohibited in certain areas. This presents a problem for cupola emission control since evaporative cooling is almost universally used as some form of gas temperature control. Heavy concentrations of water vapor exist in the effluent regardless of whether the system approach is wet or dry.

Control of opacity in an effluent is necessarily of a circumventive nature. No reduction of material emissions will influence visibility. Essentially, the methods that can be employed are either to disperse the effluent through broadside exits on dry systems or to artificially introduce sufficient temperature into the moist effluents for wet collectors. Condenser approaches are not technologically feasible because the mechanism cannot be designed for changing effluent and ambient air conditions.

Current practices are to use exhaust exits of a type shown in Figure 42 for dry collectors. If the exhaust from such a unit were concentrated in a stack and exited at the same temperature, it would be quite visible.

Raising of stack gas temperature for wet plumes is feasible where available energy resources permit. This involves a considerable amount of heat input since a large safety factor is necessary for various weather conditions. There are currently some attempts being made to utilize heat generated from secondary combustion in the upper cupola stack. This type of approach requires design of heat exchangers through which the heat is transferred to the wet gases.

SUMMARY

Control of emissions from iron foundries is basically dependent on the characteristics of the particular process involved. The cupola has been and will continue to be a problem area. There are many conditions that can influence the application of available technology to cupola emissions. These center around the nature of process conditions for a given cupola, available resources, and the predominant emission of concern. It has been the usual approach to place emphasis on control of one type of emission at the expense of control over others.

REFERENCES

1. U.S. Department of HEW, PHS, CPEHS, NAPCA, AP-51, *Control Techniques for Particulate Air Pollutants*, pp. 5-8 (January 1969).
2. A. T. Kearney & Company, EPA, NAPCA, DPCE, *Systems Analysis of Emissions and Emissions Control in the Iron Foundry Industry*, Contract No. CPA 22-69-106, December 1970.

3. Engels, G. *The Nature and Characteristics of Cupola Emissions,* Gray and Ductile Iron Founders Society, Inc., February 1969, 53 pp.
4. Engels, G., and E. Weber. *Cupola Emission Control,* Gray and Ductile Iron Founders Society, Inc., May 1969, p. 60.
5. Cowen, P. S., "Cupola Collection Systems", *Proceedings of the Total Environmental Control Conference,* American Foundrymen's Society, November 16–19, 1970, p. II-5-1.
6. Shaw, F. M., "Sulfur in Cupola Stack Gases," Journal of Research and Development Report No. 451, *6,* 444-454 (December 1956), Report No. 451, British Cast Iron Research Association.
7. "Combustibility of Cupola Top Gases with Various Humidity Contents and Temperatures," *Giesserei 52,* 197-200 (July 1965).
8. Allen, G. L., F. H. Viets, and L. C. McCabe. "Control of Metallurgical and Mineral Dusts and Fumes in Los Angeles County, California," *Bureau of Mines Information Circular 7627,* April 1952, pp. 39–48.
9. McCabe, L. C., A. H. Rose, W. J. Hamming, and F. H. Viets. "Dust and Fume Standards," Industrial and Engineering Chemistry 4, 2388-2390 (November 1949).
10. U.S. Department of HEW, PHS, CPEHS, NAPCA, AP-49, *Air Quality Criteria for Particulate Matter,* p. 12-9 (January 1969).

16.
PARTICULATE EMISSION CONTROL IN THE STEEL INDUSTRY

B. A. Steiner

Senior Environmental Engineer
Armco Steel Corporation
Middleton, Ohio

The steel industry is generally considered a dirty industry when compared with other manufacturing operations. It is indeed true that particulate emissions from uncontrolled steel mills represent rather spectacular examples of gross air pollution due to the nature and quantity of these emissions. The control of particulate emissions has been the steel industry's major effort in recent years, however, and much of the progress in alleviating these emissions has been equally as impressive as the columns of red smoke which plagued steel mills in the past.

The manufacture of the greatest percentage of steel in this country has been undertaken at large integrated steel mills which begin with the basic raw materials of steelmaking—iron ore, coal, and limestone—and carry out the necessary operations leading to the finished steel product. In general, the potential sources of air pollution from such integrated mills can be classified into 6 major categories: (1) storage and handling of raw materials, (2) coking, (3) sintering or pelletizing, (4) ironmaking, (5) steelmaking, and (6) rolling and processing.

The problems associated with each of these operations will be discussed, along with some of the experiences in controlling particulate emissions from these sources. Some steel industry trends leading to possible changes in character of emissions and approaches to control will also be discussed. Because of the complexity and scope of this subject, only an overview is possible in this text.

STORAGE AND HANDLING OF RAW MATERIALS

Iron ore, limestone, and coal are received by barge, rail, or truck and are unloaded to open stockpiles. Usually these materials are presized at the source. From the stockpiles, the material must be transferred to the process by some means.

The exposed stockpiles represent potential sources of air pollution in dry, windy weather. A number of simple spraying and wetting techniques have been employed but with only modest success. Because of the size of these stockpiles, enclosing or covering is impractical. A better means of controlling these types of emissions may lie in more efficient sizing of the raw materials and in finding better methods of handling so that the percentage of fines, which is the portion that may tend to become airborne, will be reduced.

Each time these raw materials are handled and transferred from the carrier to the process, dust is generated at the transfer point. Because these points of transfer can generally be localized, proper hooding and ventilation can control these sources. Mechanical dust collectors and fabric filters have been widely used for these purposes.

In general, the problems associated with materials handling in the steel industry are not unique to the industry, and methods of control employed in other industries are similar.

COKING

Coke is the major source of fuel used in blast furnaces and is a necessary commodity in the manufacture of steel. Coke is made by heating coal to a high temperature over a period of 10 to 20 hours to drive off the volatile matter but retain certain physical and chemical properties. This process takes place in coke ovens consisting of slots of complex silica brickwork arranged with flues for efficient heating. Each slot may be up to 20 feet high, 55 feet long, and 20 inches wide. A battery of ovens may consist of 80 to 100 such slots.

After the coal is blended and weighed, a hopper-equipped vehicle operating on top of the oven is loaded with a charge of coal. This vehicle, called a larry car, moves to the oven to be charged, and the coal is loaded through charging holes at the top of the oven. Following the heating or coking process, the doors on each end of the oven are removed, and the coke is pushed with a ram from the oven into a coke car. The car is then transferred to a quench tower where water is sprayed onto the hot coke to cool it for further transport, use, or preparation. Gases evolved during the coking process are withdrawn to a by-product recovery system for recovery of tars, oils, napthalene, etc. The cleaned coke gas is then used as fuel for underfiring the ovens and other purposes.

The three major sources of particulate emissions in the coking process are charging, pushing, and quenching. When the coal is charged to the hot oven, the gases in the oven are displaced, and the coal immediately begins to volatilize. Particulate matter is emitted through the charging holes in large rushes of smoke at this time. When the oven is pushed, dust and smoke are also emitted as the hot coke falls into the coke car. During quenching, the rising rush of vaporized water will carry fine coke particles into the air. It has been estimated that of the total particulate emissions from coke plants, charging, pushing, and quenching represent 60, 30, and 10% of the problem, respectively.

Numerous attempts have been made to control coke plant particulate emissions. Fair success has been realized in controlling quenching emissions by installing baffles or mist eliminators in the quench towers. Since the particulate is contained in the water vapor, providing surfaces for condensation of the water serves to reduce particulate emissions.

Because the charging emissions represent the major coke plant problem, most of the control effort has been in this area. One method of control has been to try to remove the smoke and gases through the collector main leading to the by-products plant. This has been largely ineffective because sufficient draft cannot be provided to overcome the tendency of the gases to be discharged at the charging ports. Stronger drafts have the tendency to create fires within the ovens. A modification of this method, which utilizes sequenced charging in combination with precise automatic timing and instrumented operations, holds more promise, however. Under the sponsorship of the AISI and with federal grant supports, a coke plant is presently being completely modified for utilizing this method. Experience and operating data at this facility, on its completion, will provide much needed technology on this approach.

Direct collection and gas cleaning of the gases discharging at the charging ports have been attempted at many installations. This method consists of shrouding areas where smoke is emitted, drawing the gases into a collection device, combusting the gases, and cleaning the gases by wet scrubbing. All of this equipment must be mounted on the larry car. Problems associated with this approach have been with design defects, maintenance, and inability to sustain proper operations. In general, this approach has not been successful.

Perhaps the most promising method for coke oven charging control is the concept of pipeline charging. This method abandons the conventional charging technique and utilizes a closed pipeline for charging the oven with preheated coal. Any displaced gases are recovered in the by-product system. and performance data will provide the entire industry with a breakthrough on control of these emissions. Added production capacity through the use of preheated coal is a side benefit of this charging method.

Emission control from pushing coke ovens has not received the attention charging has. Methods of control are only now being conceived. The major problem in pushing is the so-called green coke push, when an oven is pushed before it is thoroughly coked. Fires are started in the coke car, and large quantities of smoke are emitted into the air. Control of these instances is a problem of proper operating procedures, however, and can be avoided at well-run coke plants. Emissions from normal pushes are difficult to control because the source is difficult to contain. Several preliminary designs for closed or hooded pushing operations have been devised but none applied at full scale. The most promising of these designs appears to be those where the pushing and quenching operation are combined on mobile equipment moving the length of the battery. Gases from both are cleaned with wet scrubbers before discharge to the atmosphere.

Control of coke plant emissions in general is the biggest air pollution problem remaining for the steel industry. Although some of the proposed methods hold promise for new coking facilities, adaptation of these approaches to existing coke plants is a major concern, especially in view of the long life (40 years) of coking facilities.

SINTERING OR PELLETIZING

Sintering and pelletizing processes are intended to convert iron-bearing fine materials into agglomerates of sufficient size for charging into a blast furnace. A sinter plant consists of a central collecting facility for dusts, sludges, slag fines, and ore fines, where these materials are combined in proper proportion with limestone and coke. The mixture is ignited with a gas flame on a traveling hearth called a sinter strand. As the strand moves along, a draft is pulled through the hearth layer, and the material burns from the top down, the coke serving as fuel to sustain combustion. The fused material at the discharge end of the sinter machine is then crushed, cooled, and screened for use in the blast furnace. All fines generated are returned to the process.

Air pollutants are derived from two major sources. The first is the induced draft air which will pick up fine material as it passes through the sinter mixture. The second is a multitude of materials handling points at storage bins, conveyors, crushers, screens, etc., throughout the plant.

The induced draft discharges of sinter plants historically have been equipped with mechanical collectors and, in some cases, electrostatic precipitators. Cyclones have been installed for purposes of diminishing fan wear and do not remove particulate matter to the extent required for air pollution control. Precipitators have generally been adequate for production of the so-called acid or low basicity sinter. However, the trend toward high basicity sinter production has changed the electrical properties of the dust and has rendered precipitators inadequate by today's standards of control. Various types of scrubbers and fabric filters are now under investigation for sinter plant applications.

One factor complicating the technology of control for sinter plants is the presence of condensable hydrocarbons in the discharge gases. These appear to be vaporized from various materials in the sinter mix. The fact that these condensables can be considered particulates by the definitions generally being adopted means that they too must be removed, though removal may be more difficult. However, on the other hand, the removal of this type of material in collection devices may lead to maintenance and operating problems. These factors, in combination with the need of some sinter plants to remove gaseous pollutants and the questionable ability of precipitators to cope with high basicity sinter dusts, would appear to indicate a trend toward the use of wet scrubbers for control of sinter plant induced draft emissions.

The materials handling dust problems at sinter plants can be controlled with proper hooding, ventilation, and dust collection equipment. Fabric filters receive widespread use, and low-energy wet scrubbers have also been used. Although control techniques are available, the problem for new plants often lies in defining where the dust problem will occur.

Sinter plants have not been built in this country since the 1950s, but a number are under consideration today. With the increased quantities of dusts and sludges being produced by air and water pollution control facilities, increased efforts can be expected to recycle this material to the steelmaking process with the aid of sintering plants.

IRONMAKING

The subject of ironmaking in integrated steel mills is largely confined to the blast furnace, the purpose of which is to reduce iron ore. Ore, coke, and limestone is charged at the top of the furnace through a series of seals. Preheated air is forced into the furnace, reacts with the coke to form a reducing gas, and is discharged at the top. Iron and slags are tapped from the furnace.

Large quantities of dust are discharged with the blast furnace gas. However, because of the heating value of blast furnace gas, it is used as fuel and must be cleaned before doing so. Gas cleaning of blast furnace gas has been practiced for decades and has been the basis for gas cleaning technology being applied elsewhere today. Mechanical collectors, wet scrubbers, and electrostatic precipitators, usually in series, are used for this purpose. Two-stage scrubbers are also being used more as furnace top pressures increase and thereby increase the pressure available for scrubber operation.

Particulate matter from blast furnace operation has been and continues to be excellently controlled. Minor problems continue to exist with the tapping of iron from the furnace and casting. The formation of kish, a graphite particle which emerges from cooling molten metal, is troublesome, but the problem of controls lies not in the removal of the matter from a gas stream but with the capture of the particles at the source.

STEELMAKING

Steel today is produced in open hearth, electric, and basic oxygen furnaces (BOF). It is these operations that have created the majority of the air pollution problems in the steel industry and that have received the most attention. Each of these furnaces, by virtue of melting and refining operations, give rise to the evolution of large quantities of fine iron oxide particles. In recent years, the use of oxygen lancing to decarburize the metal has led to greater quantities of emitted pollutants.

The open hearth furnace is charged with scrap steel, fired by direct flame to melt the charge, then charged again with scrap or molten metal and refined. A heat lasts a period of 6 to 12 hours with variations in the level and nature of particulate emissions. Control of open hearth emissions has been accomplished by the use of electrostatic precipitators and high-energy wet scrubbers. Waste heat boilers are often employed in conjunction with precipitators. Scrubbers are less expensive initially but have higher operating costs. Despite the higher operating cost, scrubbers have the advantage of greater reliability and the ability to remove gaseous pollutants which can be an additional problem at some shops.

Because of the decline in use of open hearth furnaces in favor of electrics and BOFs, it is probably safe to say no new open hearths will be constructed. Also, the equipping with air pollution control devices of those existing open hearths which will be kept in operation has probably already been started or completed. Thus, design of control facilities for open hearths is a thing of the past. The manufacture of steel in open hearths was 87% in 1960 and is estimated at 20% by 1980.

The basic oxygen process accounted for only 3% of the total steel production in 1960 but has overtaken the open hearth and is expected to produce 55% by 1980. In the BOF, scrap and molten metal is charged to a large pear-shaped vessel and is injected with high rates of oxygen. Heats can be completed in less than an hour. Because of the recent entry of the process, pollution control has been included with new installations. Approximately half of the existing installations have been equipped with electrostatic precipitators and half with high-energy wet scrubbers. However, the trend is toward more scrubbers, especially in light of the recent success of the OG system. This Japanese system consists of minimizing the infiltration of air into the furnace gases by providing a movable hood which fits over the mouth of the vessel. The uncombusted furnace gases are then cleaned at the minimum volume, thereby making the gas cleaning equipment smaller in size. The cleaned gases are then flared or collected for fuel. Use of a precipitator with an uncombusted gas would not be possible because of the explosion hazard.

Electric furnaces are also increasing in importance, with capacity increasing from 8% in 1960 to an estimated 25% in 1980. This increase is due largely to the ability of electric furnaces to process 100% cold scrap. As a result, a number of minimills, consisting of electric furnace shops and rolling facilities, are being built in areas where scrap is expected to be in great supply at low costs. Electric furnaces are cylindrical, refractory-lined shells into which the charge is dropped. An arc is imparted to the charge through an electrode system, and the electrical energy melts the scrap. Oxygen is injected for refining. Evacuation systems are provided for collecting and cleaning the dust-laden furnace gases. Most installations have been equipped with fabric filters and several have applied wet scrubbers. Costs and other factors usually determine the equipment selection. Electrostatic precipitators have been tried but are generally believed a poor application for electric furnaces.

Some lesser problems are evidenced in steelmaking shops. The transfer of molten metal gives rise to the evolution of iron oxide and kish particles. Charging and tapping of furnaces creates a similar problem, as does teeming, the transfer of molten steel to ingot molds. Some efforts for control have been successful in minimizing these particulate problems when an operation can be localized; however, the problem of capturing the gases is often physically restrictive. The advent of continuous casting and its more widespread use has and will diminish the minor problem associated with teeming. Continuous casting is a process for converting molten steel into shapes for rolling, thereby bypassing the ingot and reheat steps.

ROLLING AND PROCESSING

Problems with particulate emissions from rolling and processing are limited to only a few operations. One of these is scarfing. Scarfing consists of the removal of surface defects from slabs and other steel shapes by directing an oxygen-acetylene at the steel surface. This operation actually volatilizes the steel into a fine iron oxide fume which creates dense red plumes during scarfing. The particles from this source are exceedingly fine and difficult to collect and are contained in a saturated gas stream. High-energy wet scrubbers have been used almost exclusively in this country, whereas Canadian steel companies have installed wet electrostatic precipitators in large numbers. Both appear to do a satisfactory job. A few dry electrostatic precipitator applications have been unsuccessful, with severe corrosion problems having been experienced.

A problem similar to scarfing emissions occurs on the finishing stands of high-speed hot strip mills, used for rolling steel slabs into strip. The pressures and temperature exerted create a fine iron oxide fume which has been handled identically to scarfer particulate emissions.

Pickling and cold rolling have been found to create mist and vapor problems. In the case of pickling, which consists of passing the steel strip through an acid bath for surface preparation, acid mists and vapors are evolved. These are generally collected and removed with fume scrubbers consisting of packed towers of various designs. On cold rolling mills, oils are used to serve as a coolant in the rolling process. The high pressures and temperatures on these mills tend to vaporize a portion of this oil, which is commonly collected and removed by simple mist eliminators.

Although there are a host of other finishing operations in the steel industry, most represent few, if any, problems with particulate emissions.

SUMMARY

The steel industry has perhaps the greatest potential for particulate emissions because of the nature of its operations. The industry has, however, made significant strides in coping with these problems, particularly in the area of furnace emission control. With many of its problems, technology has been developed for control to the extent required today. The outstanding exception is the control of coking operations. However, extensive development work is under way on several approaches to the problems of coke plant emission control. One very important concept to be watched closely is the development and success of the direct reduction process. This operation would replace the function of the blast furnace, but without the requirement for coke. Widespread application of this process could, therefore, substantially reduce the need to produce coke.

The basic problem encountered time after time in attempting to control particulate emissions in the steel industry is that it is essentially a series of batch or cyclic operations. For this reason, containment of the source is often difficult, particularly when transferring products or materials from one operation to the next. Hence, the answers to many of the remaining pollution problems in the steel industry may lie in the trend toward more continuous operation. We have already seen this trend emerging with the advent of continuous strip pickling and rolling, continuous casting, and the more recent emphasis on developing continuous coking processes. There are many examples where technological advances in the industry have served to diminish or eliminate pollution problems. Among these examples are the trend toward screened blast furnace burdens, the replacement of beehive coke ovens with by-product ovens, the dominance of basic oxygen furnaces over the use of open hearths, and the development of pipeline charging for coke ovens. Breakthroughs of this type will continue to provide us with solutions to our air pollution problems.

17.
BRASS AND BRONZE EMISSION CONTROL

Robert T. Hash

Product and Application Engineer
American Air Filter Company
Birmingham, Alabama

Before discussing the actual air pollution problem in a brass and bronze foundry, the characteristics of these metals should be mentioned. Brass is generally considered to be a copper-base alloy with zinc, and bronze a copper-base alloy with tin. The remelting of merely pure copper and bronze does not create as great an air pollution problem as with the other alloys since only small amounts of metal are volatized. This is due to the high boiling points of copper and tin (above 4000° F) and their low normal pouring temperatures of about 2000 to 2200° F. With this type of alloy and good melting practice, a total emission to the air should not exceed 0.5% of the process weight.

On the other hand, the brasses that contain 15 to 40% zinc are poured at a temperature near their boiling points—about 2200° F—and some vaporization of zinc is inevitable. Emissions from this type of operation may vary from about 0.5 to 6% or more of the total metal charge and 2 to 15% of zinc content through fuming, depending on the composition of the alloy, the type of furnace used, and the melting practice.

AIR POLLUTION

The air pollution problem from brass and bronze foundries can be divided into two areas: the pollution from (1) the remelt furnaces (usually gas or oil fired), and (2) the foundry sand handling system.

Air contaminants emitted from brass and bronze furnaces consist of products of combustion from the fuel, particulate matter in the form of dust, and metallic fumes. The particulate matter comprising the dust and fume load varies according to the fuel, alloy composition, melting temperature, type of furnace, and many operating factors. In addition to ordinary solid particulate matter such as fly ash, carbon, and mechanically produced dust, the furnace emissions generally contain fumes resulting from condensation and oxidation of more volatile elements including zinc, lead, and others.

As mentioned previously, the big air pollution problem generates from the zinc-rich alloys which are poured at approximately 1900 to 2000° F. This temperature is only slightly below their boiling points, therefore fractions of

171

high-zinc alloys usually boil or flash to zinc oxide. The zinc oxide formed is submicron in size, and its escape to the atmosphere can be prevented only by collecting the fumes with highly efficient air pollution control equipment.

Perhaps one of the best ways to explain the difficulty of controlling the zinc oxide fumes from the brass and bronze furnaces is to consider the physical characteristics of the fumes generated. The particle size of zinc oxide fumes is generally considered to vary from 0.03 to 0.3 microns (Figure 43). (The term micron should be defined as a unit of length that is one millionth of a meter or approximately 1/25,000 of an inch.) The zinc oxide fumes, being very small, require high efficiency control devices. Because of the large number of the small particles generated and the large total surface area of these particles, they produce very opaque effluents. Since they produce a maximum scattering of light, the effluent look dense and bad from an air pollution standpoint.

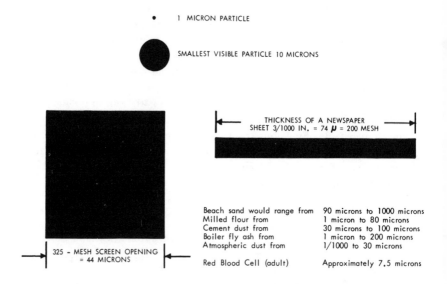

Figure 43. Particle Size Comparisons

Investigations have proven conclusively that the most troublesome fumes consist of the particles of zinc and lead compounds that are submicron in size. Oxides of these two metals can account for as much as 98% of the particulate matter in the furnace stack, depending upon the composition of the alloy.

Air or gas usage in the process gives rise to pollution. In the brass and bronze foundry, this usage, coupled with fumes from the furnace, causes

pollutant emissions. There are four principle factors which contribute to
metallic oxide fumes in brass and bronze remelting furnaces.

1. *Alloy composition.* The rate of loss of zinc is approximately proportional to the zinc percentage in the alloy.

2. *Furnace type.* Direct fired furnaces such as open hearth reverberatory
produce higher fume concentrations than crucible or electric furnaces because
the hot, high-velocity combustion gases come directly into contact with the
metal, resulting in excessive oxidation.

3. *Excessive emission results from poor foundry practices* such as improper combustion, overheating of charge, addition of zinc at maximum furnace temperature, flame impingement on charge, heating charge too rapidly,
insufficient flux cover, superheating metal, poor control of furnace atmospheres.

4. *Pouring temperature.* For a given percentage of zinc, an increase in
temperature of 100° F increases the rate of zinc loss about three times.

Careful consideration of these factors will greatly reduce air contamination.

Regardless of the efficiency of the control device, air pollution control is
not complete unless all the fumes generated by the furnace are captured.
There are many, many different types of furnaces in this industry, and we do
not have sufficient time to discuss the hooding requirements of each one. The
following should give some idea of the types of hooding that can be used and
are required to capture all fume generated.

1. *Stationary crucible furnace* (Figure 44). For this type of furnace, a
stationary, totally enclosed hood can be designed which makes a very economical type of system since the air volume can be reduced substantially. This
furnace is usually directly vented, through which venting the products of
combustion and metallic fumes are carried on to the dust collector. Auxiliary
hoods are required over the charge door and tap hole. The auxiliary hood and
the furnace vent may be tied in together so the hot combustion gases may be
diluted before entering the dust collecting system, or the auxiliary vents may
be vented to a smaller secondary dust collector.

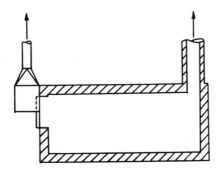

Figure 44. Stationary Reverberatory Furnace

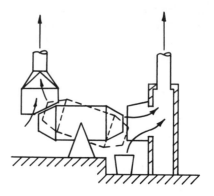

Figure 45. Rotary Tilting Reverberatory Furnace

2. *Rotary tilting reverberatory furnace* (Figure 45). The rotary tilting furnace not only tilts for charging and pouring but rotates during the melting period to improve heat transfer. Because the furnace both tilts and rotates, a direct-connection hood or control device is not practical. The furnace is under positive pressure throughout the heat, and fumes are emitted through all furnace openings. This type of furnace is probably the most difficult brass furnace to control. Therefore, the hooding requires some detailed design in order to both capture the fume and allow for the movement of the furnace.

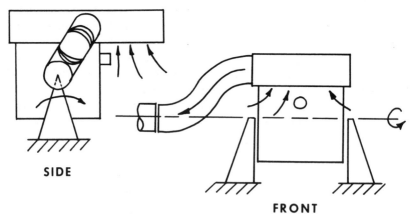

SIDE

FRONT

Figure 46. Tilting Crucible Furnace

3. *Tilting crucible furnace* (Figure 46). The crucible furnace is probably the most useful for melting small quantities of brass and bronze. The crucible has a cylindrical, steel surface shell lined with factory material inside of which the crucible is mounted. The furnace uses indirect firing, heating the crucible

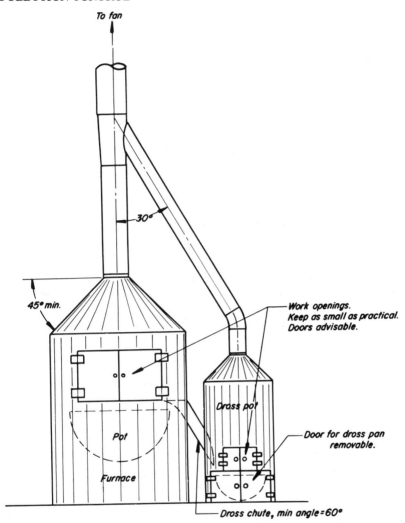

To fan

30°

45° min.

Work openings.
Keep as small as practical.
Doors advisable.

Dross pot

Door for dross pan
removable.

Pot

Furnace

Dross chute, min angle=60°

STATIONARY FURNACE OR MELTING POT

Figure 47. Stationary Furnace or Melting Pot (From Industrial Ventilation
        Manual)

contents without contacting the charge and thus reducing the amount of
emission. One kind of crucible furnace is the rotary filtering type. A hood,
such as the one shown in Figure 46, can be directly connected to the top of
the furnace and designed so that both the fumes generated during the normal
furnace operation and the fumes created during pouring can be collected.

   4. *Stationary crucible furnace* (Figure 47). A stationary, totally enclosed

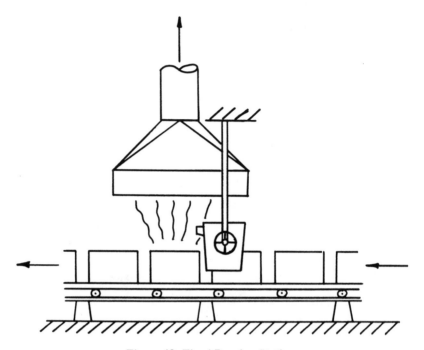

Figure 48. Fixed Pouring Station

hood can be designed, for this type of furnace, which makes a very economical type of system since the air volume can be reduced substantially over that of a canopy hood.

The emissions that are generated from the pouring stations where the molten metal is poured from the ladle into the molds is also an emission problem. These can generally be hooded in one of two ways: (1) a fixed pouring station where the molds travel on a conveyor past one pouring point can be hooded with a fixed canopy hood (Figure 48), and (2) a pouring ladle with traveling fume hoods can be used where the ladle must move from point to point to pour into stationary molds (Figure 49).

DUST COLLECTORS

Dust collectors that are normally used for both the furnace and the pouring station are fabric collectors. The reverberatory furnace, because of its direct evacuation and high flue temperature (somewhere around 2000° F), has probably the most elaborate dust collection system. Figure 50 shows a water-jacketed flue which is designed to cool the entering air temperature from 2000° F to approximately 900° F. The water that drains from the

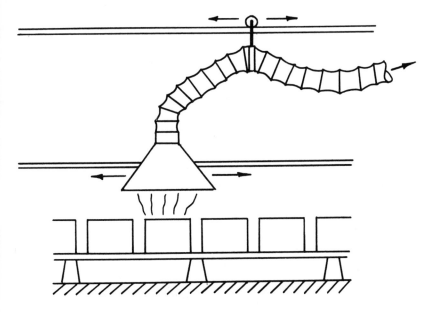

Figure 49. Pouring Ladle with Traveling Fume Hood

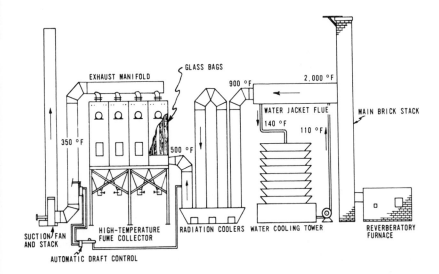

Figure 50. Gas Cooling for Bag Filter Gases (From Air Pollution
          Engineering Manual)

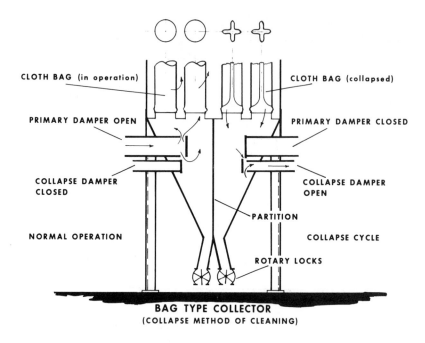

BAG TYPE COLLECTOR
(COLLAPSE METHOD OF CLEANING)

Figure 51. Bag Collector, Collapse Method of Cleaning

water-jacketed flue must be cooled and then recirculated back to the flue. The 900° F air then enters radiation coolers that cool the airstream from 900° F to approximately 500° F, the temperature required to protect the glass fabric filter. A fabric collector which uses glass bags is usually categorized as a collapse unit because the bags are cleaned by reverse air flow which actually collapses the bags, thus knocking the dust cake from the filter medium into the hopper (Figure 51). This type of collector is also considered to be an intermittent collector because it cannot operate continuously unless it is compartmented with one or more collectors. While the bags are being cleaned, the unit or that section of the unit being cleaned must be shut down (Figure 52). One of the big advantages of this type of collector, besides withstanding the high temperatures, is that on very large baghouses—for example, 30-foot bags—the collapse type of cleaning is the least expensive. The glass medium has one disadvantage in that the periodic cleaning of the bags gradually breaks the glass fiber, thus causing a higher maintenance cost.

In many cases, especially on the other types of furnaces where the indraft of ambient air is sufficient enough to cool the exhaust gas below 275° F, other media such as orlon and dacron have been used. This offers the advantage of a stronger medium which can be cleaned by different methods. One such method is used in the shaker fabric collector (Figure 53). The dust-

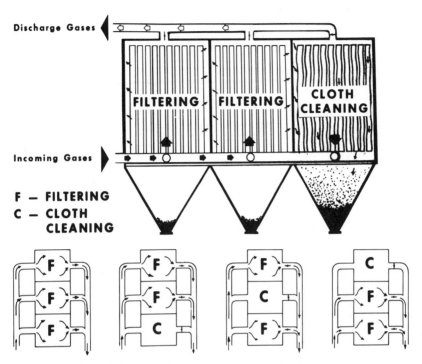

Figure 52. Bag Filter Comparents and Cleaning Cycle

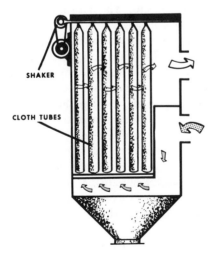

Figure 53. Shaker Collector

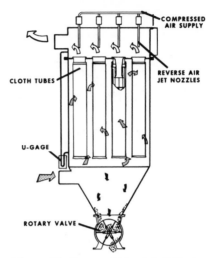

Figure 54. Reverse Pulse Jet Collector

ladened air enters the collector and immediately hits a baffle sheet which directs the air directly down into the hopper where the larger particles will fall out and the finer particles will travel to and collect on the inside of the fabric media.

The shaker collector, like the collapse collector, is an intermittent collector, and when the bags are shaken, the unit or compartment of the unit must be closed off so that no air flows through it while the bags are being cleaned. The air-to-cloth ratio, which is the air volume per square foot of cloth medium or the velocity through the media, is approximately 2.5 or 3:1 for the collapse and the shaker collector. The normal pressure drop for both is approximately 5 to 6 inches water gauge.

A newer fabric collector is the reverse-pulse jet collector (Figure 54). The dust-ladened air enters the bottom of the collector, and the dust is collected on the outside of the bags. The bags are cleaned by compressed air at the top of the unit. The usual operation is to pulse a row of bags with a quick shot of compressed air, which actually shocks the bags into releasing the dust cake, and the dust falls into the hopper. The big advantage of this type of collector is that it is a continuous fabric collector, meaning that it does not have to be shut down in order to clean the bags. It also offers the advantage of higher air-to-cloth ratios in the ranges of 8 or 9:1, which makes this unit much smaller than the collapse or shaker collector. A felted medium such as dacron felt should be used, and it is a little more expensive than the bags that are used in the shaker collector. The first cost of the pulse jet is a little more than that of a shaker collector.

The fabric collector is probably the most economical dust collecting equipment used on the usual brass and bronze melting operations, but a wet

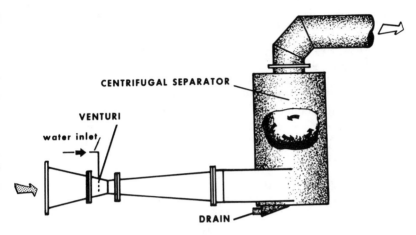

Figure 55. Venturi Scrubber

collector should be considered in particular cases where dirty, oily scrap is fed into the furnace and the oil content in the exhaust system to the fabric collector may be high enough to blind the bags, which would cause a considerable amount of difficulty with the baghouse operation. In this type of operation, the only wet collector which will give the cleaning efficiency required to meet today's more stringent air pollution codes is the high-energy Venturi scrubber (Figure 55). This collector uses a Venturi-shaped throat section where the gas velocity reaches 9000 to 24,000 feet per minute (which is in the neighborhood of 100 to 280 miles per hour) in the smallest part of the throat area. Water introduced ahead of the throat section is atomized by the high-velocity gas stream, and the dust particles collide with and are captured by the millions of small droplets. After the water is injected into the throat, it must be eliminated from the gas stream; therefore, the water-ladened gas stream enters the water separator tangentially, where the droplets are removed by centrifugal force and impingement. The droplet-free gas then passes through the separator outlet, and the slurry is continuously drained from the water eliminator section.

The pressure drop across the Venturi throat may be in the neighborhood of 50 to 60 inches of water to meet most existing air pollution codes. The water rate to the throat area will be approximately 8 to 12 gallons per thousand cfm of gas. The materials of construction of a wet scrubber depend on the fuel that is used to heat the furnace and the type of fluxes that are used during the operation of the furnace. Of course, the more exotic the materials of construction of the wet scrubber, the higher the cost of the collecting system. The following lists some advantages and disadvantages of the fabric versus wet scrubber.

FABRIC COLLECTORS

Advantages	Disadvantages
1. High efficiency.	1. First cost high.
2. Recovers product dry.	2. Maintenance cost high.
3. Pressure drop and horse-power requirements low.	3. Bags may be damaged by overheating.
4. No water pollution exists.	4. Condensation will produce caking and interfere with operation.

WET SCRUBBERS

Advantages	Disadvantages
1. Tolerates high temperature.	1. May create water disposal problem.
2. Can collect gases as well as particulates.	2. Product is collected wet.
3. First cost low.	3. Corrosion problems are more severe than with dry systems.
4. Maintenance cost is relatively low.	4. Steam plume opacity may be objectionable.
5. There is no condensation problem.	5. Pressure drop and horsepower requirements high.

The foundry sand handling equipment can be defined as a system consisting of a device for separating the casting from the sand mold and the equipment that is required to recondition the sand. The separating device is usually a mechanical vibrating grate called a shakeout. The shakeout system, as well as the conveyor transfer points and handling points in transporting the sand from one area of the foundry to the other, must be hooded and exhausted to dust collecting equipment. In most nonferrous foundries, there is a low metal-to-sand ratio, thus the bulk of the sand remains damp during shakeout, and the dust concentrations that are exhausted are low as compared to a steel foundry which makes larger castings with higher metal-to-sand ratios. Because of the wetness of the sand, a wet collector rather than a fabric collector is usually used on foundry sand. If a fabric collector is used, special consideration must be given to the possibility of water condensation on the bags, thus causing blinding.

A medium-efficiency wet collector will meet most of today's air pollution codes on foundry sand handling. A medium-efficiency wet collector is defined as a collector which has a pressure drop of approximately five to six inches water gauge.

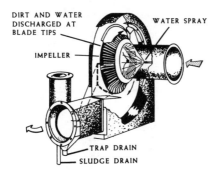

Figure 56. Wet Dynamic Precipitator

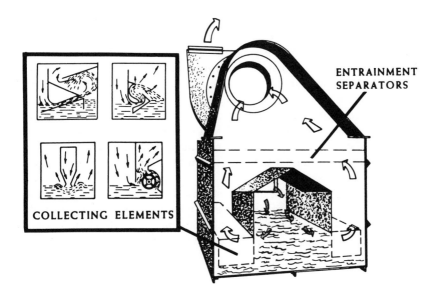

Figure 57. Wet Collector, Orifice Type

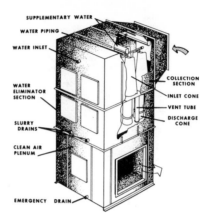

SUPPLEMENTARY WATER

WATER PIPING

WATER INLET

COLLECTION SECTION

WATER ELIMINATOR SECTION

INLET CONE

VENT TUBE

DISCHARGE CONE

SLURRY DRAINS

CLEAN AIR PLENUM

EMERGENCY DRAIN

Figure 58. Cyclonic Scrubber

The wet dynamic precipitator (Figure 56), is a medium-efficiency collector and fan combination. Dust-ladened air is drawn into the inlet, where the particulate is subjected to water spray. The water is conveyed to specially designed blades, and a fine film of water continuously flows from the center to the tips of the blades. Dust is precipitated by centrifugal force into the flowing water surface, and the water and entrapped particulate is conveyed to a drain chute. The cleaned air flows through the center of the blades into the clean air outlet. The water required is low, ranging from ½ to ¾ gallons per 1000 cfm.

The orifice type wet collector uses many different kinds of collecting elements, ranging from perforated plates to specially shaped orifices (Figure 57). Some use a bath of water that is continuously recirculated by the pump action of the air. The total water usage is in the range of ½ to 1 gallon per 1000 cfm.

There are a number of different cyclonic collectors available, ranging from simple wet cyclones to more sophisticated types (Figure 58). These cyclonic collectors utilize centrifugal force to bring the scrubbing liquid into contact with the dust. The water rate is in the range of 3 to 4 gallons per 1000 cfm.

Section IV

Raw Material Processing and Air Pollution

18.
DUST CONTROL TECHNOLOGY AND THE CRUSHED STONE PRODUCER

Michael J. Natale

The Johnson-March Corporation
Philadelphia, Pennsylvania

Several methods have been used to control atmospheric emissions from stone quarrying. One, the wet dust suppression method, has proven to be the most economical and the most effective overall.

Wet dust suppression is the utilization of a liquid for the control, reduction, or elimination of airborne dust or for the suppression of such dust at its source. Effective control involves:

1. The confinement of the dust within the dust-producing area by a curtain of moisture.
2. The wetting of the dust by direct contact between the particles and droplets of moisture.
3. The formation of agglomerates too heavy to remain airborne or too heavy to become airborne by combining small dust particles with each other and with the droplets of liquid.

From the standpoint of economy, availability, and safety, water is the ideal fluid. From the standpoint of effectiveness, however, it leaves much to be desired. As a result, a technique has been devised in which very small percentages of specially formulated dust control compounds are blended with water to greatly enhance its dust suppression qualities. The terms "dust control compounds," "wetting agents," and "surface active agents" are frequently used interchangeably. Although dust control compounds represent a specific type of wetting agent, a great many wetting agents are far from satisfactory for dust control. "Wetting agents" is a broad category which covers such items as emulsifiers, solubilizers, detergents, foamers, penetrants, thickeners, etc. Dust control compounds, on the other hand, are carefully formulated blends in which one or more special surface active agents have been incorporated.

How these dust control compounds perform can best be explained by a brief discussion of surface active agents and surface activity. There is, in all liquids, an attractive force between molecules. These forces vary inversely as the sixth power of the distance between molecules. Thus, in the liquid phase where molecules are crowded together, the intermolecular forces are powerful, whereas in the liquid-vapor phase where the molecules are relatively far apart, these forces diminish very rapidly. In the body of a fluid, the attractive forces are equal in all directions, but at the surface there is an uneven distribution of forces resulting in an inward pull. This force with which the surface molecules are attracted inwardly is called "surface tension." This term is usually reserved for those situations where we measure the interface between a liquid and air or between a liquid and its vapor. Where we deal with the force which keeps two surfaces apart or causes them to coalesce, we use the term "interfacial tension."

Fluids on high surface tension exhibit poor wetting, spreading, and penetrating qualities. Water has the unusually high surface tension of 72.75 dynes per cm at 20° C.

To reduce this surface tension and correct other defects of plain water, we turn to the use of the special surface active agents referred to above. These materials are synthetic organic chemicals which, when present in solution in fairly small concentrations, exhibit striking surface and interfacial phenomena. Although such materials may differ greatly in chemical composition, they have a characteristic feature in common: their molecules are composed of two groups exhibiting differing solubility characteristics. One part, usually a long-chain hydrophylic or water-loving group, is usually a sulfate, sulfonate, hydroxide, ethylene oxide, etc.

When properly formulated dust control compounds are introduced into water, the chemical molecules, because of their unique configuration, tend to migrate to and concentrate in the surface layer.

There, by exhibiting weaker force fields, they overcome the tendency of the water to present a minimum surface and effect an appreciable reduction in surface tension. The more effective dust control agents have been found to reduce this value to as low as 27 to 29 dynes per cm.

What does this mean in terms of efficiency?

1. *Dust can now penetrate into a droplet rather than merely coat its surface.* Plain water, because of its high surface tension, presents a tough surface which limits dust particles to the exterior of the droplet. With treated water, the entire volume of the droplet is available for dust collection.

2. *Fine particles, because they are more readily wetted, now are cemented to each other and encapsulated by the surface active droplet.* As a result, as much as a 500-fold increase in dust capturing ability is realized.

3. *The aggregates, because of increased weight, either drop from the air stream or are unable to become airborne.*

4. *Treated water, because of low surface tension, is more readily atomized:* more droplets are produced per unit volume, large increases in surface area are obtained, and contact potential with dust particles vastly improved. (Sprays of untreated water yield large droplets with low surface area and large, wasted volumes.)

5. *Treated water can wet faster, spread further, and penetrate more deeply and uniformly.* Moisture is more efficiently utilized and less moisture is required to do the job.

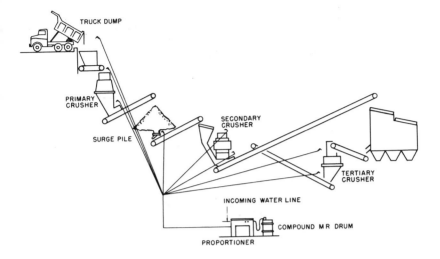

Figure 59. Application of a Dust Suppression System in a Typical Rock Crushing Plant

It should be noted that, if the dust control compound is properly applied, agglomeration of the dust particles will not contaminate the stone because only a minimum amount of moisture is required and screening will remove the dust. Plain water applied to the stone tends to muddy its surface and produce a coating which is undesirable.

Wet dust suppression is a dust suppression method which uses moisture in the form of water and a dust control compound. This is applied to the stone or ore to condition the material so that it becomes dust free in the material handling operations. It is not the purpose of a wet dust suppression system to fog a dust source with a fine mist to remove the airborne dust, but to wet the dust and to cause the dust particles, after wetting, to agglomerate. It is, therefore, primarily a dust suppression system with the object of preventing dust from becoming airborne rather than of removing the dust from the air or of causing it to precipitate from the air after it has become airborne.

This method of dust control is particularly suitable to rock ore crushing plants since it does not involve providing crushers, screens, conveyor trans- fers, etc., with hoods or other enclosures. The equipment remains open and accessible for maintenance while the dust normally created at these points is suppressed. It allows the operators to readily see all equipment and material flow in all open areas.

Since the dust is not collected, there is no secondary disposal problem as may occur if the dust is handled dry, and no water pollution problem as when the dust is handled wet.

Figures 60 and 61. Before and after photos of Typical Application of Dust Suppression System in a Small Quarry

The effect of the moisture applied to the material produces a carryover dust control effect to points beyond the actual moisture application locations.

This provides dust control throughout the entire handling system and suppresses the dust in open areas of the plant where dust collecting systems are not possible or practical, such as at the discharge of conveyors to surge piles, stock piles, or truck or rail car loading.

In general, applications must begin as soon as possible in the materials handling system. This is important since the time required for proper distribution of moisture on the stone or ore is a critical factor in achieving dust control. Applications normally begin at the truck or car dump at the primary crusher. This introduces moisture to the material and suppresses the dust created during the dumping operation (Figures 60 and 61).

As expected, applications are normally made at all crushers where new dry surfaces are created and where it is possible to utilize the action of the crusher to assist in mixing the moisture into the new dry surfaces of the stone. Application, therefore, would normally be made at the primary crusher, at the secondary crusher, and at all tertiary or recrushers. It can be readily appreciated that, even though the stone coming to the crusher is properly and uniformly treated and in a dust-free condition, the new dry surfaces created by the fracturing of the stone in the crushers will require additional moisture.

In addition to these applications, treatment is normally made at feeders under surge or at reclaim piles if the stone remains on the piles for such a length of time that evaporation necessitates the reapplication of moisture to return the stone to its dust-free condition.

Applications are not normally required at conveyors, transfers, screens, conveyor discharges to stock piles, or to surge piles since the stone, previously properly conditioned, can be handled at these points without dust.

The wet dust suppression method has been used effectively on a wide variety of stone and ore such as limestone, traprock, granite, shale, copper ore, iron ore, taconite, nickel ore, dolomite, asbestos, sand, and gravel. It can be generally considered to have a universal application to stone or ore handled through a normal crushing and screening plant.

Through proper control of moisture, the mesh sizes can still be readily screened from the larger sizes without being affected by the wet dust suppression applications.

Before design of a wet dust suppression method can be considered, it is necessary to make an engineering analysis of the plant requirements. First, the plant should be surveyed in detail by a qualified engineer to determine the operating conditions and the nature of the problem. Some of the factors which must be determined are:

1. type of material being handled;
2. material flow through the plant;

3. general plant layout including overall dimensions;
4. size and type of crushers, conveyors, screens, bins, feeders, and other equipment;
5. retention time of material in bins or on stock piles;
6. tonnages at various points;
7. material temperatures at various points;
8. type of water available for the system;
9. electrical characteristics available;
10. areas requiring protection against freezing; and
11. major dust points and conditions which occur at these points during normal operation.

Initial costs for a wet suppression system will vary, depending upon the capacity and complexities of the materials handling facilities, the availability of electric and water services, the physical distances involved, and prevailing local labor conditions. For these reasons, it is impossible to formulate standard cost figures in terms of dollars per ton, number of application points, or some other norm.

To give some concept of cost, the price of an average 500-tons-per-hour plant would be approximately $20,000 for a complete, turn-key wet dust suppression system. This would, therefore, be the complete cost for equipment, engineering, field erection, and electrical wiring.

Because of small horsepower requirements and low water consumption, the cost of power and water is generally negligible. The principal operating cost is the wetting compound.

Our experience has shown that, based on a full year's operation because of varying weather conditions, the average annual cost of treatment of a ton of stone or ore is approximately one mill per ton.

There are some types of operation where the use of a wet dust suppression system is not practical for all areas of the materials handling facility. The application of a wetting compound is not recommended, for example, to hydrated lime, cement, or other moisture-sensitive materials.

In high-temperature material conditions, evaporation of the moisture and cooling of the material will occur, which must be compensated for in the design of the dust system.

Where dust is created by a product consisting entirely of fine particles, such as from ball and rod mills, the relatively enormous surface areas and large amounts of moisture required do not lend themselves to a standard wet dust suppression system. For such applications, special conditioning equipment can be utilized which is specifically designed to permit controlled application of the wetting solution to a controlled materials flow. This makes possible the conditioning of fine materials for dust control without affecting the normal handling characteristics of such fines. It must be borne in mind, however, that because of the vast surface areas with which we are dealing, the required amount of moisture for adequate dust control is much higher than with normal stone products, and a great care must be taken to determine whether this fine product can stand the 4 or 5% moisture normally required.

On rare occasions, situations have occurred whereby a standard dust suppression system became troublesome because of wet stone being processed or because of other conditions not usually found in a quarry. The problem encountered under such conditions is usually the coating of stone, resulting in a marginally salable product. In this situation, a combination wet suppression and dry collection system offers the most practical and effective solution to the dust control problem. This is so because a combination system permits the benefits of the wet suppression system at the screening, transfer, and store-out points and at truck loading stations. The dry collecting benefits are gained at the crushers, where dust is created by extracting most of the fines which would otherwise adhere to the salable product, and collecting it into bins.

Experience in the industry has indicated that a dry collection system in itself cannot be designed to provide dust control at all the required dust control points at a reasonable economic cost. We are, therefore, now going to project for you a typical stone plant utilizing a combination wet and dry suppression system.

As has been outlined above, a wet dust suppression system can provide effective dust control in many different types of material handling systems and can enable the plant to comply with both local and state air pollution control requirements. While this method can generally be employed to obtain the required degree of air pollution control in most rock crushing plants, it is often necessary or advantageous to incorporate additional methods of dust control. A combination of a wet dust suppression system and a dust collector will, in certain situations, provide more desirable results than either system can if used separately.

For example, a secondary crusher is a major source of fine dust. If the stone is conditioned by a wet suppression system prior to entry to the crusher, we will have dust control at the crusher inlet. At the same time, with an air evacuation system in operation at this area, most of the fine particulate normally produced at the discharge of the crusher will be collected.

After this step, a very light treatment of dust suppression compound will produce a dustless condition at subsequent transfer points, screens, bin discharges, and stockpiles without fear of stone contamination. This is an important fact to note, since experience has taught us that, even after the materials have been subjected to a draw-off system at the normal application points, there are still sufficient fines in the stone to produce a dusty condition at screens, stockpiles, etc.

A combination wet suppression and air evacuation system is recommended for the following types of applications:

1. Where the fine particulates have an economic value and product recovery could be intrinsically profitable in addition to meeting local air pollution control requirements.

2. If the secondary plant production requirements make it necessary to utilize very fine screens, a combination system would be less troublesome by preventing blinding of the screens—because of moisture—by a large percentage of the -325-mesh dust particles which are normally the cause of screen blinding.

The average rock crushing plant poses a special problem in air pollution control because of its physical extent and the number of crushers, screens, conveyor transfer points, feeders, and storage and reclaim facilities (Figure 59). This extensive materials handling system with its multiplicity of dust-creating points makes it impractical to rely solely on an air evacuation system.

To use an air evacuation system as the only means of control would require a large central collector with an extensive ductwork system or a number of individual collectors located at strategic points throughout the plant. Dust would normally become airborne at crushers, screens, conveyor transfers, storage bins, etc., and could be collected at such points. However, there would be no carryover effect such as is obtained with a wet dust suppression system so that there would still be a serious dust problem at the screens, stacking out, or bin loading areas and windage loss from conveyors and stockpiles.

On the other hand, a combination system offers the following advantages:

1. The load on the air evacuation system is greatly reduced so that smaller collectors with smaller fans and motors can be utilized.
2. The wet dust suppression system will require fewer application points and less compound per ton; consequently, costs will be reduced.
3. Any problem of stone coating will be eliminated.
4. Optimum dust control efficiency will make it possible to meet local and state code requirements in all areas of plant operation.
5. Where product recovery is profitable, installation can pay for itself and become an economic asset.

There is no easy, quick, or simple formula which can be applied to the air pollution control problems in rock crushing plants. Each plant presents its own unique problems in terms of the materials being handled, the types of materials handling equipment used, plant layout and size, tonnages to be treated, local conditions, and air pollution control regulations.

It is essential, therefore, that all these variable factors be carefully analyzed by competent and experienced engineers so that the system furnished will provide the required control while keeping the capital expenditure to the minimum.

19.
PARTICULATE EMISSION CONTROLS IN PORTLAND CEMENT MANUFACTURING

J. L. Gilliland

Technical Director
Ideal Cement Company
Denver, Colorado

In the United States, portland cement is manufactured at about 175 plants, which normally are located near the principal markets or the source of raw materials. They may be the major industry in a rural area or small city, or a minor part of the industrial complex in a large city. The principal air pollution sources are the kiln and the clinker cooler, both of which are controlled by electrostatic precipitators or high-temperature fabric bag collectors.

PORTLAND CEMENT MANUFACTURING PROCESSES

The raw material for portland cement is about 80% calcium carbonate in the form of limestone, marl, or shell. The minor raw materials are sources of silica, alumina, and iron oxide, such as clay, shale, sand, and iron ore.

Portland cement is manufactured by the dry or the wet process, which refers to the method for preparing the raw materials and feeding the kiln. In the dry process, the raw materials may be dried before being introduced into the grinding mill or they may be dried within the raw grinding circuit. Rotary driers are used for predrying. When the dry-grind system is used, hot air from the kiln or from a separate furnace is introduced into the grinding circuit. In the wet process, the raw materials are ground with 30 to 45% water to give a pumpable slurry.

Dry process kilns may be fed directly with the blended, finely ground raw material or the feed may be introduced into preheaters, which utilize the waste heat in the kiln gases to preheat the feed before it enters the rotary kiln. Wet process kilns are fed directly with slurry, which is dried as it starts its passage through the kiln.

Portland cement kilns rotate at 60 to 90 rph. They are direct-fired with gas, coal, oil, or combinations of these fuels. The gaseous discharge consists of the products of combustion, water vapor, carbon dioxide, and nitrogen as well as the water vapor and carbon dioxide which are driven off from the raw material. Particulates in the gaseous discharge are partially decarbonated raw material enriched by the sodium and potassium salts volatilized from the raw material. When coal is used as fuel, the coal ash combines with the raw material, supplying some of the siliceous component.

The product discharged from the kiln is called clinker. It drops to a cooler, which serves to preheat the combustion air for the kiln. Cooling air in excess of the kiln requirements is discharged to the atmosphere.

The clinker is then conveyed to the finish grinding section, ball or tube mills close-circuited with air separators. The finished cement is conveyed to the cement storage silos, usually by pneumatic conveyors.

AIR POLLUTION CONTROL

Bag collectors are widely used to control particulate emissions from crushing and grinding equipment and material handling operations. A typical modern cement plant will utilize 30 to 50 bag collectors. Grinding mill circuits are usually controlled by bag collectors although electrostatic precipitators (ESP) have been installed on some finish mill circuits. When the kiln gases are used for raw material drying, the raw mill circuit is usually controlled by an electrostatic precipitator.

The principal particulate emission sources are the kiln stack and cooler stack, as illustrated in Figure 62. Formerly, cyclone collectors were considered adequate for both applications because they prevented nuisance dustfall conditions. Now, however, opacity and discharge weight rate restrictions require more efficient collection equipment.

Particulates are removed from kiln gases by electrostatic precipitators or woven glass fabric bag collectors, either of which may be preceded by cyclone collectors. Scrubbers have had very little application because of the problems in handling particulates which react with water. The choice between the use of an ESP or a bag collector is based on several factors.

The gaseous discharge from a dry process kiln contains insufficient moisture for satisfactory ESP operation unless it has been used to dry the raw material. Its temperature may be too high, in which case the gases must first be cooled by air dilution, radiation from cooling loops, or humidification. After humidification, dry process kiln gases can be controlled by ESP. Bag collectors may be preferred if the gas cooling has been accomplished by air dilution (which increases the gas volume requiring treatment) or by radiation. Figure 63 is a dry process plant equipped with a pressure-type Fiberglas bag collector. Note the absence of a stack.

For wet process kilns, the ESP has had general acceptance. Figure 64 is an ESP unit on wet process cement kilns. In some cases, bag collectors have been satisfactory; but, in others, bag collectors have resulted in severe operating problems. Either ESP units or bag collectors must be well insulated to prevent moisture condensation. High alkali salt concentrations may result in fabric blinding, greatly increasing the pressure drop through a bag collector.

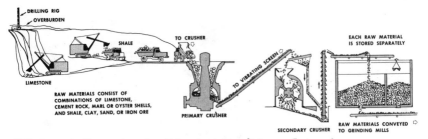

1 Stone is first reduced to 5-in. size, then ¾ in., and stored

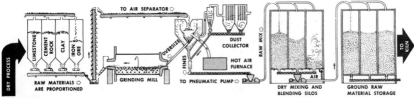

OR **2** Raw materials are ground to powder and blended, or

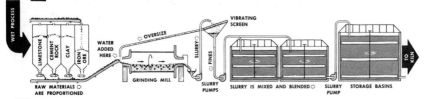

2 Raw materials are ground, mixed with water to form slurry, and blended

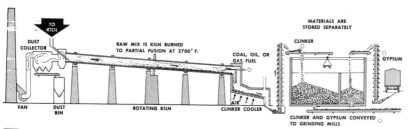

3 Burning changes raw mix chemically into cement clinker

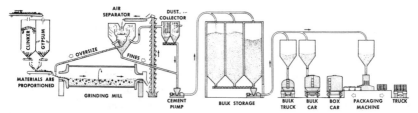

4 Clinker with gypsum added is ground into portland cement and shipped

Figure 62. Steps in the Manufacture of Portland Cement
Courtesy Portland Cement Assoc., Skokie, Illinois.

Figure 63. Dry Process Kiln Controlled by Pressure Bag Collector

The use of multistage ESP units affords a generalized example of the increasing cost-benefit ratios which result when extremely high collection efficiencies are required. In a recent installation, four stages were used, each designed to have an efficiency of 84%. If the collected dust in the first stage is assigned a cost of unity per ton of material collected, cost of material collected by the succeeding stages is as follows:

Stage	Overall Efficiency	Relative Cost Per Ton
1	84.00%	1.00
2	97.44%	6.25
3	99.59%	39.01
4	99.94%	242.31

Three types of collectors are being used on the exhaust air from air-quenching clinker coolers. The first and most popular is the bag collector, which may use fabric of Fiberglas or of Nomex®, a high-temperature nylon. Figure 65 shows a pair of Nomex fabric collectors on an air-quenching cooler. ESP units have been installed on a few clinker coolers. The third type uses a gravel-bed filter which is cleaned by backwashing with atmospheric air.

Figure 64. Electrostatic Precipitator Controlling a Wet Process Kiln

Under normal operation, the temperature of the exhaust air from the clinker cooler presents no problem. However, when a surge of hot material is discharged from the kiln, the air temperatures may exceed the allowable operating limits for Fiberglas (550° F) or Nomex (425°F). Emergency water sprays can be installed to handle these upset conditions. Because of the possibility of a thermal overload which would destroy the bags, gravel-bed filters are finding increasing application to the clinker coolers on very large kilns.

The clinker cooler exhaust air contains little moisture, so ESP installations are provided with water sprays for gas conditioning. Provisions for good evaporation are essential because free water will combine with the clinker dust to form a material buildup.

Figure 65. Bag Collectors on Clinker Cooler Exhaust

EMISSION CONTROL REGULATIONS

Legal restrictions based on "equivalent opacity" have been generally applied to portland cement kiln emissions. However, manufacturers of ESP units will not guarantee performance to a visual standard. It is their consensus that an exit loading of 0.015 gr/ACF will result in a "clear or essentially clear" stack. For a wet process kiln, large quantities of water vapor are discharged in the stack gases with the result that a heavy steam plume appears when the atmospheric temperatures are low and the humidities high.

Emission limitations based on process weight have been adopted by many state and local jurisdictions. In calculating the allowable emissions, the feed rate rather than the product weight is used. In terms of barrels (376 lb) of production, the dry raw feed is approximately 600 lb, and the slurry feed is approximately 1000 lb. Production is no longer measured in barrels. In terms of short tons (2000 lb) of clinker, the dry raw feed is 1.65 tons and the slurry weight is 2.74 tons per ton of clinker produced. When coal is the fuel, the weight of the coal is added to the raw material weight to arrive at the process weight.

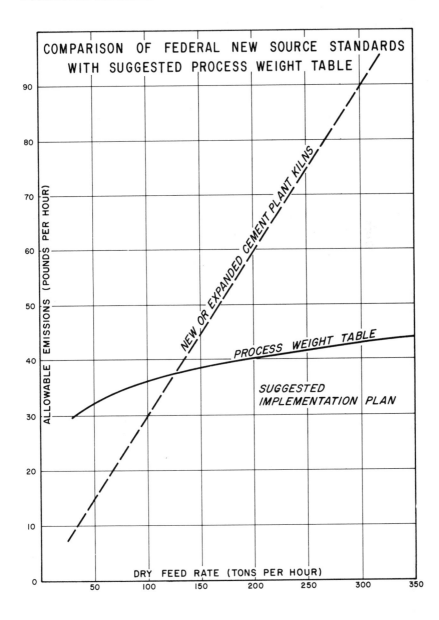

Figure 66. New federal source standards are less severe than the older method represented by the process weight table.

The federal "new source standards of performance" for portland cement plants were published in the Federal Register for December 23, 1971, and appear in 40 CFR 60. These standards limit the visual emissions to 10% equivalent opacity. Weight-rate particulate emissions are limited at the kiln stack to 0.3 lb per ton of dry raw feed to the kiln. Emissions from the clinker cooler stack are limited to 0.1 lb per ton of dry raw feed to the kiln. Performance tests are based on the dry portion of the EPA sampling train. Collection efficiencies well above 99.9% are usually required to meet these regulations.

Emission limitations based on process weight tables are not consistent with the new-source standards. For smaller process weights, the new source standards are more severe, as would be expected, inasmuch as they are intended to reflect the "best demonstrated technology." For large sources, the emission limitations based on the process weight table become increasingly more severe than the new source standards. This is illustrated in Figure 66, which utilizes the process weight table from Appendix B, 40 CRF 51.

20.
CONTROL OF ATMOSPHERIC EMISSIONS FROM CONCRETE BATCH PLANTS

Robert T. Hash

Product and Application Engineer
American Air Filter Company
Birmingham, Alabama

The purpose of concrete batching plants is to store, convey, measure, and discharge the constituents for making concrete to transportation equipment. There are generally considered to be three types of plants: (1) wet batching, (2) dry batching, and (3) central mix.

The wet batching plant mixes sand, aggregate, cement, and water in the proper proportions and dumps the mix into transit mix trucks which mix the batch en route to the site where the concrete is to be poured. Dry batching plants mix the sand, aggregate, and cement and dump into flat bed trucks which transport the batch to paving machines where water is added and mixing takes place at the job site. The central mix plant uses a central mixer to dump the wet concrete into open bed dump trucks where it is delivered to the job site by this means.

WET CONCRETE BATCHING PLANTS

Wet concrete batching plants are probably the most common type of plant throughout the United States. The typical wet concrete batching plant elevates sand and aggregates by belt conveyor or bucket elevator to overhead storage bins or silos. Cement from bottom discharge trucks or railroad cars is usually conveyed to an elevated storage silo.

Sand and aggregates of a batch are weighed in a weigh hopper by additions of the material from the overhead bins, and cement usually is delivered from a screw conveyor from the silo to a separate weigh hopper. The weighed aggregate and cement are then dumped into a common hopper which drops the mix into the receiving hopper to the transit mix truck. At the same time the mixture is dumped into the transit mix truck, the required amount of water is injected into the flowing stream of solids. Dust, the air pollution contaminant, results from the handling of the material that is used. The problem areas of wet concrete batching plants are generally:

1. unloading of cement from trucks or railroad cars and storing it in silos;
2. venting of displaced air in the cement weigh hopper during batching;

3. emissions of the particulate matter with the displaced air from the rear of the truck mixer drum during charging into the mixer;

4. subsequent handling of the cement dust that is collected by the air pollution device; and

5. general handling of the aggregates—hoppers, bins, screw conveyors, elevators, and pneumatic conveying equipment.

The typical cement receiving storage system is at or below ground level (Figure 67). If the hopper is designed to fit a canvas discharge tube from the hopper of the truck or railroad car, very little dust should be emitted at this point. The dust is transported from the receiving hopper, usually through an enclosed bucket type elevator, to the cement silo. The cement silo must be vented to allow the air displaced by the cement to escape. Unless this escape point is filtered, a significant amount of dust will be emitted, causing an air pollution problem.

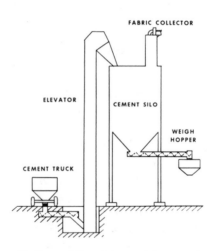

Figure 67. Cement Receiving and Storage System

The silo can be vented to a central dust-collecting system or to a single collector placed on top of each silo. A fabric collector is most often used to vent the cement silo as well as other dust-collecting points in concrete batching plants. The single filter placed on each silo can be operated without an exhauster when the material is delivered to the silo by bucket elevators because it is simply used to filter the air that is forced out.

Many concrete batching plants now receive cement from trucks equipped with compressors and pneumatic delivery tubes. In these plants, a single filter vent used for the gravity filling of cement has proved inadequate, and other methods of control are required. In the pneumatic delivery system, the volume of conveying air is approximately 350 to 700 cfm, depending on the loading cycle, etc. Since the air is being forced into the silo, the baghouse will require a blower in order to relieve the pressure inside the silo. A normal fan capacity is in the neighborhood of 1200 to 1300 cfm. Therefore, the silo is actually under a partial negative pressure, or vacuum, during the filling operation. This is very important because it prevents cement dust leakage around access doors, etc., which would not be the case if the cement silo was under a positive pressure during filling.

The cement weigh hopper may be a compartment in the aggregate weigh hopper or it may be separate. Since the weigh hopper is filled at a rapid rate and the displaced air entrains a significant amount of dust, the dust that is generated may be controlled by venting the displaced air back to the cement silo or by venting it directly to a central collecting system.

Another point of emission of dust to the atmosphere is when the weighed amount of sand, cement, and rock is dumped from the weigh hoppers into the receiving hopper of the transit mix truck. There are different ways of collecting the dust generated at this point, but the control of the dust depends on a proper hood design. Listed below are three methods of controlling at this point.

1. A telescopic shroud which encompasses the rear of the mixer and which can be controlled mechanically or by air cylinders. The flexible shroud is lowered over the rear of the truck when the truck is in place. (Figure 68).

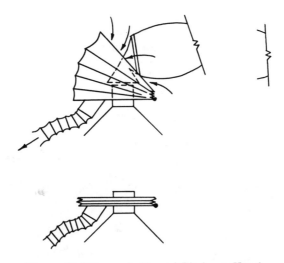

Figure 68. Telescopic Shroud Discharge Hood

2. A stationary hood that the trucks must back into. This is a good design, but the fact that the trucks have to back into it may be a disadvantage. (Figure 69.)

FRONT VIEW

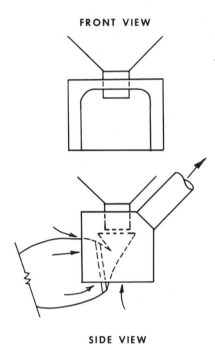

SIDE VIEW

Figure 69. Stationary Discharge Hood

3. A sheet-metal hood which totally encloses the transit mix truck receiving hopper when it is in place. After the truck is filled, the panels are raised. The side panels of the hood are actuated by air cylinders. (Figure 70.)

Figure 70. Sheet-metal Hood

DRY CONCRETE BATCHING PLANTS

Dry concrete batching plants have been used mostly where large amounts of concrete are needed, for example, road construction work. Because cement, sand, and rock are dumped into trucks in a dry state, the dry batching process poses a much more difficult air pollution control problem than the wet batching plants. Most plants that do dry batching also do wet batching, and, since the receiving hopper of the usual transit mix trucks is several feet higher than the top of the flatbed truck used to haul the material in dry batching, the weigh hoppers must be set high enough to accommodate the transit mix trucks used in wet batching. The result is a long free fall of material when a dry batch is dropped. This, of course, produces a considerable amount of dust.

Because the plant operator must see the operation and because the truck must also have freedom of movement, this is a difficult operation to hood. The truck bed is usually divided into several compartments and batch is dropped into each compartment. This necessitates moving the truck after the drop into one compartment so another compartment is in place. A canopy hood just large enough to cover one compartment at a time provides effective dust pickup and affords adequate visibility (Figure 71). The sides can be made of heavy rubber to give the hood some flexibility so that it will not be damaged if a truck hits it. The hood is sometimes mounted on rails to permit it to be withdrawn and allow wet batching into transit mix trucks. The exhaust volume, of course, varies with the shape of the hood but is usually in the neighborhood of 6000 to 7000 cfm.

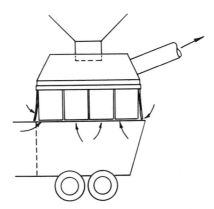

Figure 71. Dry Batching Discharge Hood

CENTRAL MIX PLANTS

As far as the cement silo and general material handling is concerned, the central mix plant has the same general ventilation as do the other two types of plants described, but, in a central batch operation, concrete is mixed in a stationary mixer and is discharged into a dump truck in a wet state.

The air pollution problem that differs from the other two types of concrete batching plants is brought about when the dry aggregate and cement falls rapidly into the large mixer. The consequent rapid displacement of air creates a considerable amount of entrained dust and air.

Effective control of the dust generated at the discharge end of the mixer is a function of good hood design—one flexible enough to be moved out of the way when the mixer is dumped—and adequate ventilation air. One such hood design is a hydraulically operated swingaway cone-shaped hood (Figure 72). For a hood of this type, the indraft face velocity should be approximately 1000 to 1500 feet per minute in order to overcome the high velocities that are created when the dry aggregate and cement fall into the mixer.

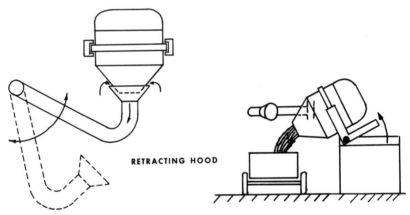

RETRACTING HOOD

Figure 72. Central Mixer and Retracting Hood

AIR POLLUTION EQUIPMENT

The dust collector that is normally used on concrete batch plants is of the fabric type. The fabric collector is one of the oldest types of collectors and employs the unique principle of using the dust itself to collect other dust. The dust forms a cake that bridges across the interstices of the fabric medium. The different types of fabric collectors are categorized by the different ways in which the dust is cleaned from the cloth media. Generally, two types of fabric collectors are used on concrete batching plants: (1) the shaker and (2) the pulse jet.

The Shaker Collector

The dust-laden air usually enters the shaker collector (see Figure 53, page 179) through a baffled plenum which provides even air distribution and maximum protection against abrasion of bags and metal surfaces. A change of direction and reduction of inlet velocity within the plenum causes heavy particles to fall directly into the hopper below.

The finer particles travel upward into the cloth filter tubes where the dust is trapped on the inside surface of the fabric. Clean air passes through the cloth to the clean air plenum and is discharged to the atmosphere. The collected dust is periodically discharged by a shaker mechanism and falls into the hopper. The shaker collector is an intermittent collector because, when the bags are being cleaned, the unit or a compartment of the unit must be shut down. In order to operate the shaker collector on a continuous basis, a compartmented collector must be used.

Continuous duty units are arranged in two to four compartments, each having a shaker mechanism. Inlet ducts are connected to a common manifold joining each compartment. Compartments are periodically isolated by individual automatic dampers.

Dacron or Orlon bags are recommended, and an air-to-cloth ratio of approximately 3:1 should be used.

The Pulse Jet Collector

As opposed to the shaker collector where the dust is collected on the inside of the bag, the dust is collected on the outside of the pulse jet collector (see Figure 54, page 180). Clean filtered air passes through the inside of the tubes through the clean air plenum and on to the atmosphere. At preset intervals, a wired controller energizes solenoid valves, allowing compressed air to flow through a pipe orifice, and air moves down the tube by gravity and inertial force. The tube is deflected rapidly, "pulsing" the medium first downward and then upward. This action causes the collected dust cake to break loose and fall into the hopper.

The complete cleaning cycle is continuous and automatic. The pulsations occur with the fan running and without the need of compartmented housing, dampers, and secondary blowers. Higher air-to-cloth ratios of approximately 10 to 12:1 can be used with polypropylene felted bags.

21.
AIR POLLUTION CONTROL FROM HOT MIX ASPHALT PLANTS

Fred Kloiber

Director of Environmental Control
National Asphalt Pavement Association
Riverdale, Maryland

Asphalt hot mix plants produce a thermoplastic mixture of heated mineral, natural or synthetic aggregates, and asphalt cement that is placed on a prepared road base or subbase and compacted. After cooling, its thermoplastic properties terminate and a more or less flexible road structure is ready to provide a strong and smooth surface for safe and dust-free traffic.

The hot mix batch plant is more functional than decorative and is designed to produce between 100 and 500 tons of hot mix per hour.

The principal particle emission source is the rotary dryer, not unlike the kiln in portland cement manufacture. Large quantities of air, between 12,000 and 80,000 cfm, are required to support combustion and to carry away the moisture liberated from the mineral aggregate.

The fuels used in drying these aggregates range from natural and L.P. gas over light fuel oils to Bunker "C" oil.

Other points of emission are the hot elevator that transports the heated aggregate to the mixing tower in which the aggregate is separated into four or more sizes. The aggregate is then proportioned into a weigh hopper and from there into a twin-shafted pugmill. Asphalt cement is added, and, after a short mixing period, the finished product is dropped through a gate into delivery trucks.

Another less often applied principle is that of a plant where aggregate is fed by volume into a continuous mixer and discharged over an adjustable dam into a discharge hopper.

The dust carry-out from the rotary dryer contributes by far the highest percentage of emissions from a hot mix plant. Although traces of unburned fuel particles and the normal materials produced when petroleum fuels are burned (sulphur oxides, carbon oxides and nitrogen oxides) are emitted, they are not considered to be significant by any of the control agencies.

The mineral particles in the range of 0 to 100 microns are the main concern in the environmental control of a hot mix asphalt plant. One governing factor is the dryer drum gas velocity. The smaller the diameter of a dryer drum, the higher the gas velocity will be and with it the dust

211

carry-out; the same holds true conversely. The bigger the diameter of the dryer drum, the lower the gas velocity will be, and the chance of particles becoming airborne will decrease.

This phenomenon is extremely important in the operation of the plant since the difference between low velocity (600 ft/min) and high velocity (900 ft/min) may increase dust carry-out by 125%.

The grain-size distribution of mineral particulates is of great importance to the type and efficiency of collection. The finest mesh size generally considered in the specifications of the hot mix industry is 200 mesh or 74 microns, although most aggregates carry particles in the submicron range. It is increasingly difficult to prevent particles from escaping through the stack as they become smaller and smaller whether these fine particles are to be collected through wet or dry systems.

Hot mix asphalt plants have a long history of emission controls. Since the transport of the hot mix constitutes a substantial portion of the product price and since excessive heat loss may render the finished mix useless, hot mix plants have to be close to the urban areas where the biggest share of the plants production is used. To eliminate complaints by neighbors, equipment ranging from knock-out boxes to multiple-cone collectors were installed. One advantage of these installations lies in the fact that valuable fine materials were collected and fed back into the hot elevator. A disadvantage was that more horsepower was necessary to overcome the static pressure of these simple collectors.

As time went on, denser population and the first air pollution control regulations mandated further reduction of emission and, with that, a higher grade of sophistication of equipment. The first spray towers, centrifugal wet collectors, and dynamic wet systems were installed. Responsible industry organizations anticipated the advent of more and more restrictive air pollution control regulations and urged their memberships to voluntarily comply with self-imposed regulations.

The National Asphalt Pavement Association urged its members as early as 1965 to adopt a voluntary stack exhaust control of 0.16 grains/scf for urban or residential areas and 0.21 grains/scf for remote areas.

When the Clean Air Act became law on the last day of 1970, the hot mix industry found itself in an entirely new ball game.

Ambient air standards were promulgated, emission standards were prepared by the states to be included in the "implementation plans" that had to be submitted to the newly created Environmental Protection Agency in January of 1972.

Emission control regulations which had been promulgated by many states in the years prior to the Clean Air Act were reappraised to insure compliance with the ambient air regulations.

Research regarding control of asphalt hot mix plants has been hampered by the fact that the mineral aggregate feed is dependent on the locally available materials—crushed limestone, basalt, granite, sandstone—many types of ore tailings and sands and gravels in a multitude of combinations with a range of 1 to 20% minus 74 micron (200 mesh U.S. Standard Sieve) materials down to the submicron range. These materials contain silt, clay, and finely ground rock dust.

While 2 to 8% of these fines constitute an average percentage of the feed to the plant, many operators have to contend with and control up to 22% of -74 micron material.

Recent studies have revealed that further grain-size distribution down to the submicronic size shows extreme variations within that low micron range.

In some rock dusts, 10% of the -74 micron material is below 5 microns; in other dusts, 50% of -5 microns is not uncommon.

To control this extremely wide range of grain-size distribution, each of the 4500 hot mix plants has to be tested and evaluated separately.

A further complication arises from the fact that each hot mix plant produces a number of different asphalt mixes, where coarse aggregate containing very low amounts of dust are combined with fine aggregates in an infinite range of combinations at a moisture content ranging from 1 to 12%. These aggregate blends require varying volumes of air to carry away liberated moisture.

Air pollution control equipment in hot mix plants consists generally of two or more stages. The primary control equipment usually separates the coarse particles from the airstream. These simple devices usually are effective only in the particle sizes down to about 100 microns and suffer a reduction of their collection efficiency rapidly in the below 100 micron size range. The most commonly used primary dust collectors in hot mix plants are:

1. *Skimmers and expansion chambers.* These simple particle collectors are based on the principle of decreasing the gas velocity by expanding the ductwork to allow airborne particles to reach a terminal velocity and settle out by gravity. To increase collection efficiency, baffles are sometimes added. The particles fall to the bottom of the device and are discharged through a valve to the material elevator.

2. *Centrifugal dry dust collectors or cyclones.* Hot mix plants employ large-diameter single collectors, several medium-sized collectors, or multiple-tube dry collectors.

All these cyclones have three common design features:

1. *A cylinder-shaped outer body* with either a shave-off slot or a cone-shaped body cut off at the bottom, which is closed off by a rotary (butterfly) valve or a flap gate.

2. *An axially centered inner tube* projecting partially into the outer body with an outlet at the opposite end.

3. *A tangential inlet* for the dirty gas through the side of the body at the center tube side.

As the particles, in a spinning motion, proceed downward to the lower part of the collector, pressed against the outer wall by centrifugal force, the cleaned gas escapes through the inner tube, the vortex, to the next stage of the collection system.

The maintenance of centrifugal collectors has been a problem in the hot mix industry. The majority of the maintenance-oriented employees at hot mix plants are not familiar with the importance of the purpose of the airlock, which prevents "false" air from entering the collector and interfering with the collection process by re-entraining already collected particles.

Large cyclone collectors, properly designed, have proven to be extremely effective in removing all particles down to 30 microns, but the design criteria call for excessive power demand.

By dividing the system into a number of cyclones ranging from two to a multitude of cyclones diminishing in size, the power demand for equal collection efficiency drops. The theoretical design of multiple-cyclone systems depends on equal distribution of the workload to all involved small cyclones. The theoretical efficiency is practically never accomplished due to differential grain-loading in the gas stream and unequal distribution of power.

The scavenger or fugitive dust system, which is an integral part of the overall control system of the modern hot mix plant, is employed to put negative pressure on all potential emission points such as the hot screens, weigh hopper, and pugmill, and sometimes on the hot elevator. The fugitive dust from these emission points is usually in the very small particle range due to the low gas velocity applied to these points. As a rule of thumb, 10% of the total airflow requirements are devoted to this system. The inlet to the general collection system is usually located in the duct before primary collector(s).

It is common to have hot mix plants equipped with two units of primary collection equipment using either two pieces of identical equipment, or any combination thereof, and always with the higher efficiency equipment downstream from the less efficient one.

Secondary systems in asphalt hot mix plants usually fall into two categories: (1) wet collection and (2) dry collection.

Electrostatic precipitators have been used very rarely to my knowledge; only one hot mix plant is controlled by such a device and, therefore, will not be covered here.

Wet secondary systems can be classified on the basis of the energy required to overcome the static resistance of these units:

Unit	Energy Requirement
Spray chamber	low
Centrifugal or cyclonic	low
Dynamic	medium

          Orifice   medium to high
          Venturi   medium to high

The spray chamber performs similarly to the expansion chamber in the dry system. Water spray is added to intercept particles and to carry them out of the bottom of the chamber in a slurry, while the cleansed air escapes through the stack.

In the wet cyclone as in the dry, the same centrifugal force is used, water being added either from the center or from the outside wall through spray nozzles to promote impingement and carry-out through the bottom of the cone.

The dynamic wet collector is, in essence, a centrifugal wet washer in which the fan impeller is sprayed with water. Thereby, the impingement of particles and the collection efficiency is increased.

The orifice washer is a device where the dust-laden gas is forced through an orifice together with water to cause impingement.

A stage control system combining the higher class centrifugal dry and wet collectors is employed in hot mix plants. Many of the existing state air pollution emission regulations may be complied with *providing* that the -30 micron material does not constitute a substantial percentage of the ingoing grain-loading. The importance of the grain-size distribution is emphasized in a study sponsored by the Environmental Protection Agency. The results show that hot mix plants with low-energy scrubbers at times had lower emission rates than plants with high-energy scrubbers. This fact can be based on the size distribution and concentration of the in-going grain-loading.

The highest state of the art is undoubtedly represented by fabric filters and Venturi scrubbers. The increasing stringency of control regulations on the federal level as well as in some of the states virtually dictate the installation of one or the other.

The fabric filter has been used with varying success over the years in the asphalt hot mix industry.

The advent of synthetic fibers in conjunction with advancing research and development has advanced technology to the point where generally successful filter collection systems can be designed for this industry. The fabric manufacturers have developed fibers and weaving processes that are highly efficient in collecting particulates and, at the same time, withstand the high temperatures that occur in hot mix plants. Much research has been done on how particles arrange themselves on the filter. The understanding of the influence of the cleaning cycle has increased. The mechanical cleaning by shaking the bag or by a cloth ring has yielded to cleaning by reversed air.

However, the design of a baghouse depends mainly on the dust-load concentration going into the collector and the grain-size distribution of the dust load. Exaggerated claims have been made by some vendors to operators of hot mix plants; guarantees have been written on baghouse performance when visual observation alone shows that these guarantees are not being complied with; and control officials have been convinced that a baghouse *per se* is the salvation from all compliance problems.

Usually, emission control regulations are combined with opacity requirements. Some of the most stringent particulate emission regulations are set below visibility requirements.

Today, if in our industry a stack does not have a visible plume, the operator is practically automatically exempt from compliance testing. Under the auspices of the Clean Air Act, the Environmental Protection Agency is at this time considering a performance standard for new and substantially modified emission sources. One of the industries affected is the asphalt hot mix industry.

The state of the art, the Environmental Protection Agency contends, is an emission control of 0.03 grains per standard cubic foot. But, of course, state of the art is dictated by the economical feasibility of the application of such a standard.

A complete baghouse system installation on today's market, according to an equipment manufacturers association study, cannot be purchased for less than $100,000, which is an additional expenditure of 30 to 35% to the cost of a new hot mix plant and is at least 50% of the value of the average existing hot mix plant.

Venturi scrubbers consist of a converging or "throat" section and a diverging section which are followed by a dewatering tank. As the dust-laden gas enters the converging section, it is accelerated to high velocities as it passes through the throat.

Water as scrubbing liquid is injected either into the converging section or directly into the throat. The high velocity of the liquid coming into the diverging section atomizes and causes impingement of the particles. An added cyclone, here called "separator," washes the impinged particles away.

The efficiency of a Ventura depends on the velocity of the gas-water combination through the throat, which is simply a function of applied power that may be continued to extremely high efficiencies. Control agencies contend that the efficiency-cost relationship for Venturis becomes uneconomical when production exceeds 150 tons per hour. It should be noted that 90% of our plants exceed this production rate.

Members of the hot mix industry and related equipment manufacturers have attacked the particulate control problem from the opposite side; *i.e.*, to prevent dust from becoming airborne in the first place. To this end, a manufacturer in Germany has developed a process where the cold and wet aggregate is blended with atomized asphalt. The material is then processed through an "activator," which is essentially a rotary dryer with an attached pugmill. While the mineral aggregate is dried and heated, the asphalt globules become fluid and coat the fine particles, thus preventing them from becoming airborne.

A similar technique, the "turbulent mass process," is being studied in this country. Water, asphalt, and admixtures are fed with the mineral aggregate into the dryer. The particles are coated with the asphalt-water mixture and thereby prevented from becoming airborne.

Although a substantial amount of dust is captured by these two methods, it cannot be expected that these processes will meet the control regulations proposed for new emission sources. They will pose a collection problem if fabric filters are used since the escaping particles will be coated with asphalt. Also, at the present time, it cannot be stated unequivocally that the mix produced in these processes will be of the same quality as the mix currently being produced. The current specification requirements eliminate the use of some of these processes and, therefore, would require a substantial modification of these specifications.

A third process is the cascade dryer, which operates on very low gas velocities and thereby allows fewer particles from becoming airborne.

The overall state of the art in the hot mix industry is such that reasonable codes, as promulgated by the majority of the states, can be met with reasonable costs. However, if the extremely stringent suggested control regulations for "new emission" sources on a nationwide basis should be promulgated, the cost of the control equipment would constitute extreme hardships for the average owner.

22.
PREVENTION OF AIR POLLUTION FROM COAL REFUSE PILES

John W. Walton

Chief, Engineering Program
Tennessee Department of Public Health
Division of Air Pollution Control
Nashville, Tennessee

A problem facing the mining industry for many years has been what to do with their coal refuse. Through the years, this coal waste was dumped in the easiest, most convenient spot and no further attention was given to it. Nothing was done about the smoke and fume emissions from these burning piles. The coal company officials felt that this was a natural phenomenon associated with mining. The general public suffered under the polluted conditions caused by these burning piles because they were unaware the problem could be eliminated. Public opinion on the problem of burning coal refuse began to change as increased emphasis was placed on air pollution and public health. The public began to demand that action be taken on eliminating or controlling the fumes and smoke from these piles.

In the United States, state-wide legislation on controlling air pollution from coal refuse piles was first contemplated in Pennsylvania in the early 60s. West Virginia and Kentucky, the other major coal-producing states, followed Pennsylvania's lead in the mid-60s. Recently, Virginia joined in regulating this problem.

The problem of burning coal refuse piles is not limited to the United States. The total extent of the problem is not known, but it appears in all coal-producing countries of the world. Great Britain and Germany have legislation on the problem, and information on the prevention of this pollution source is available from those countries.

It is difficult to imagine the conditions surrounding a burning refuse pile. The most concise and graphic description that could be found appeared in the *Louisville Courier-Journal:*

> A burning gob pile is a good earthly approximation of hell. Wisps
> of steam hiss from the earth underfoot. An occasional blue flame
> licks out of the ground. A stench of sulfur is heavy in the air, and
> yellow sulfur coats the edge of cracks in the baked earth.

Actively burning or smoldering coal refuse piles do emit smoke and fumes. Unpleasant odors are also associated with them. Various authorities disagree on the gaseous pollutants that are liberated from these burning piles. Gaseous pollutants emitted during this slow combustion process may consist of carbon dioxide, carbon monoxide, oxides of nitrogen, hydrogen sulfide, and sulfur dioxide. One investigator reported finding mercaptans and ammonia in addition to sulfur dioxide and hydrogen sulfide.

The air contaminants may produce several effects in the surrounding community. There have been reports of property and plant damage in areas adjacent to burning piles. The health of residents in the vicinity may be affected, and certainly a burning pile does not enhance the desire of a person to live in the area.

FACTORS CONTRIBUTING TO A BURNING PILE

The type of coal preparation used, specific gravity setting for mined material, and desired ash content affect the amount of coal in the refuse. A higher percentage of coal in the refuse causes the pile to be more combustible. If the pile is constructed in a haphazard manner, the chances of its igniting are greatly increased.

Most of the fires in a coal refuse pile are caused by spontaneous combustion. The tendency of coal in the refuse pile to reach a temperature for spontaneous combustion varies with the rank or type of coal mined. Studies have indicated that the lower the rank of coal the easier it is to ignite from spontaneous combustion. The tendency to ignite lessens as the quality of the coal rises.

In a coal containing a high percentage of organic sulfur, the chances of spontaneous combustion are very low. Studies have indicated that coal refuse piles containing more than 30% pyrite or more than 15% sulfur are prone to spontaneous combustion. Inherent moisture in certain ranks of coal has a definite effect in increasing spontaneous combustion.

The particle size of the coal in the refuse is also important in the possibility of spontaneous combustion. Dumping of the refuse in an indiscriminate manner may cause segregation of the different particles sizes and expose an increased amount of surface area to oxidation. Disposal of refuse by aerial tramway and by truck dumping without additional work on the refuse pile tends to enhance segregation of the refuse and accents any tendency toward self-heating within the pile.

Other factors also contribute to spontaneous combustion: erosion, pile stability, and the dumping of fresh refuse on an old pile. A frequent cause of spontaneous combustion is foreign material dumped on a refuse pile or vegetation left at the site.

Although spontaneous combustion is responsible for most of the coal refuse fires, accidental ignition plays a role in starting some fires. Accidental ignition may be caused by (1) careless burning of trash on or near the refuse pile, (2) camp fires, (3) forest or brush fires, (4) mine fires, or (5) dumping hot boiler ashes on refuse piles. The first cause—indiscriminate burning of rubbish—ranks as the most common cause of accidental ignition.

The length of time that these piles will burn depends on the amount of combustible material in the pile. The refuse pile usually contains less than 25% combustible matter such as coal, pyrite, and bony and various combinations of carbonaceous material. One U.S. Bureau of Mines survey has indicated that the average coal refuse pile could burn for twenty years.

METHODS OF REFUSE DISPOSAL

While no method exists for the absolute prevention of fires in refuse piles, except for total immersion in water, the use of sound disposal techniques can minimize the possibility of ignition. The main principle in successful disposal of coal refuse is to prevent air from entering the pile. Disposal techniques include underground stowage of refuse, various methods of surface pile construction, and returning refuse to the pit in strip mining operations.

Underground Stowage

Underground stowage or returning the refuse to the mine is practiced primarily to combat mining subsidence. This method is slow and expensive since the refuse is worked back into mined-out galleries. Most of the refuse in the Ruhr is stowed underground within 12 to 20 hours after mining. The cost of stowing averages 38 to 50 cents per ton of coal produced.

If underground stowage is not practiced, there remain only two possible methods of refuse disposal. The refuse can either be disposed in a surface pile or in strip pits. The disposal procedure chosen usually reflects the topography of the area in which the refuse is located.

Surface Disposal

The topography of the disposal area will determine the type of pile construction. The primary step in pile construction is always removing foreign combustibles from the disposal area. In hilly terrain, the disposal methods must be modified to conform to the site restrictions. In less mountainous areas where the terrain is level, refuse disposal is a relatively simple operation.

The circular pile is suggested for level terrain since the greatest volume of refuse can be contained in a smaller area. The larger the pile's diameter, the more economical it is to deposit the refuse in circular, well-compacted layers of refuse two feet in thickness. Every six feet, the refuse is sealed by a wall of inert material. Each wall is terraced to reduce erosion. Terracing also reduces the fire hazard by checking the draft along the slope of the pile. It has been suggested that the access road be built into the perimeter of the pile so that the road bed and the road (surfaced with crushed rock) forms part of the pile seal.

The size of this pile would be limited by the equipment available and whether it is more economical to add more layers or to build a new pile. The cost of the sealing material increases slightly as the height of the pile is raised. The larger the diameter of the pile, the less the cost. The overall cost of this disposal technique has been estimated at less than one cent per ton of refuse.

Mountainous terrain presents a more difficult problem for refuse disposal. Rectangular pile construction is recommended for restricted terrain in narrow valleys, and semicircular construction is recommended for wide valleys or rolling terrain. These disposal techniques offer safe disposal for less than one cent per cubic yard of refuse (about one cent per ton of refuse).

The most advantageous location for a refuse pile is a dry valley or one where the drainage can successfully be diverted around or under the pile. Such a site allows the building of the refuse pile across the valley from hill to hill, thus exposing only one face of the refuse pile to the elements.

Normally, such a valley cannot be found, and the refuse pile is located against one side of a hill. A narrow valley favors a pile of limited width while a wider valley with low slopes allows considerable latitude in pile construction. These methods leave three sides of the pile (the front and ends) exposed to the elements. In a valley with an appreciable slope, it has been suggested that the upper end of the pile should be tapered to the valley floor, leaving only the front and lower end of the pile exposed.

Compaction

Refuse is compacted in many ways. Many coal companies use bulldozers as their main method of compaction. The dozer is aided on the surface of the pile by the movement of the refuse trucks. In using an aerial tram, all the compaction obtained comes from periodic working of the refuse pile with a dozer. (Aerial tram disposal without this additional compaction is one of the worst methods of refuse disposal and should be avoided.)

If the disposal area is compacted sufficiently, 25 to 30% more refuse can be placed on the site. The starting place for the pile should be the lowest point of the hill or slope so that it can be kept level as refuse is added. Figure 73 shows the compaction of the refuse in layers from the bottom up. Refuse is brought initially to the bottom of the hill, Y, and raised through successive levels to point Z. Spreading and compaction of the refuse can then be accomplished. This figure shows that the finished pile will have a continuous slope. Terracing has been suggested to prevent excessive erosion. The terraces should be at 10-foot intervals with an 8-foot flat and a 35° slope between terraces.

This ideal method of disposal can be handled properly only during good weather. During the winter and inclement weather, an additional step must

be interposed between refuse dumping and ultimate disposal. It is a fundamental law of soil mechanics that maximum compaction cannot be
achieved if the water content of the material being deposited exceeds its
optimum moisture content.

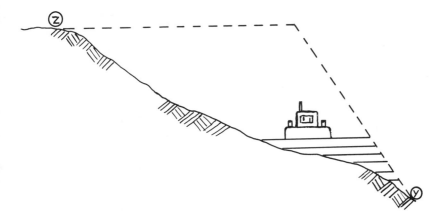

Figure 73. Compaction in Layers from the Bottom Up. (Source: Spoil Tip
Management, National Coal Board, Production Dept., England)

An alternative disposal method has been devised. In this method, the
refuse is temporarily deposited over the hillside above the disposal site. The
refuse remains there until it is drained and returns to a workable condition.
When weather permits, the refuse is undercut and moved down to the
disposal site for spreading and compaction (Figure 74).

While the refuse must be placed in compacted layers, as the pile rises
the exposed edges of the pile must also be compacted to prevent the
ingress of air. Reduction of the normal angle of repose of the refuse pile to
a maximum of about 50% is recommended.

The low slope will reduce pile erosion and normally provide access for a
bulldozer to compact the edges. It will decrease segregation of refuse since
the material is spread by the bulldozer and reduce the chances of spontaneous combustion in the pile. In the event a "hot spot" develops, this type
of slope would provide access to the problem area so it could be controlled. No matter how the pile is constructed, there is a rule that must be
followed: the sides of a pile must be sloped in such a manner as to permit
compaction of all exposed surfaces. It can be either a continuous slope or
terraced. Leaving exposed sides on the pile is one of the most common
faults of constructing a refuse pile.

Layered thickness is the primary factor in obtaining good refuse compaction. Thick layers are more likely to be honeycombed with void spaces,

Winter Disposal Area

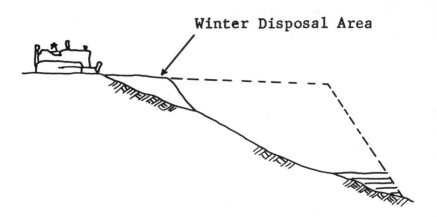

Cross Section

Winter Disposal Area

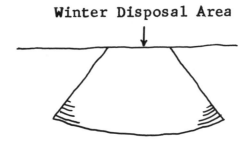

Toe

Elevation

Figure 74. Winter Disposal and Summer Spreading. (Source: Spoil Tip Management, National Coal Board, Production Dept., England)

which support combustion; thin layers tend to avoid this problem. Adequate compaction is limited to a depth of approximately two feet. The equipment used to layer the refuse will also affect the compaction obtained.

Sealing

It is not enough to compact the refuse pile; the sides and edges of the pile must also be sealed with an inert material. Earth and clay, which prevent the entrance of air into the pile, are suitable sealing materials. The seal should be at least two feet thick in ordinary circumstances, and, if the refuse has a high acid content, the seal should be four feet thick. A seal in conjunction with refuse placement by layering and compaction offers protection from spontaneous combustion or accidental ignition.

Refuse Reactivity

While compaction and sealing combined are the ideal disposal methods, it has been suggested that the disposal practices be based on the reactivity of the refuse. Refuse of low reactivity need only be layered and compacted, but more reactive refuse requires that the exposed edges be sealed with an inert material. The most reactive refuse requires a layer of inert material between refuse layers in addition to the inert seal at the edges.

Weathering

Certain clays and shales in the coal refuse spall and disintegrate into small flaky particles when exposed to the weather. It has been suggested that refuse with these weathering characteristics should be allowed to oxidize for a period of a week or two to reduce its heating tendency. During this weathering period, the particle size distribution in the refuse pile is reduced. The refuse should be allowed to weather in truck-size loads to obtain maximum dissipation of heat. This greatly enhances refuse compaction in the pile when it is ultimately layered.

While the refuse is weathering, there is a danger that acid will be formed from any pyritic content in the waste material. This could lead to acid drainage problems because of run-off from the refuse pile. Weathering requires additional disposal space so that the individual truck dumps may be oxidized to the maximum extent during this period. Since all refuse does not have weathering characteristics, this technique should be based on the individual refuse involved.

Crushing

Crushing is an infrequently used method of disposal primarily for economical reasons. The refuse is removed from the preparation plant, crushed and transported to the disposal site. The suggested size for satisfactory compaction is -3/8 inch. The cost of crushing to this size is estimated at slightly less than five cents per ton of refuse crushed.

A study on the feasibility of using this disposal method suggested that a crusher installed in a new preparation plant will reduce the overall cost of the plant. This reduction results because the larger slate and rock is crushed to a smaller size so the conveyor belt needs to be only one-half the size required to handle uncrushed material.

Further justification of the crushing method is that, when the refuse is crushed and placed in compacted layers, the chances of spontaneous combustion are reduced. The crushed refuse is easy to handle and can be contoured and rounded to permit the minimum amount of erosion. If there is insufficient sealing material available for a pile, crushing would be a solution. Many coal operators reject the procedure since it is an extra process that yields no return on their investment.

Strip Mining

The placement of refuse in strip pits is a desirable means of disposal. There is always plenty of noncombustible material at hand to seal the surfaces and edges of the refuse. It has been suggested that all exposed coal seams and outcrops in these pits be protected with an inert material before they are covered with refuse to prevent the possibility of igniting the coal seam.

Whichever technique is used, economics begins to play an important role in refuse disposal. In 1957, it was reported that $107 million was spent on the disposal of refuse in the United States. If money is to be invested in a disposal area, the life of the pile should be of prime consideration; the site selected should be capable of holding at least ten years of refuse production.

CONTROL OF BURNING COAL REFUSE PILES

Extinguishing a burning refuse pile is a difficult and costly task. Studies on the various methods used to extinguish burning coal refuse piles go back to the 1930s.

The control techniques are normally classified in five categories, as follows.

1. Digging out the fire or isolating the fire area by trenching.
2. Pumping water onto the fire area and its immediate vicinity.
3. Applying a blanket or cover of incombustible material such as limestone dust, shale dust, or slag dust over the fire area.
4. Injecting a slurry of rock dust or other incombustible material with water through drill holes into the fire area; grouting with cement.

5. Spraying water over the fire area.

The primary consideration in deciding which control technique to use is the construction of the refuse pile. Isolation appears to be the ideal solution for long and narrow piles. For blanketing to be adequately used, the refuse pile should be leveled before adding the selected material. This method requires continual surveillance since most materials tend to crack with heat and permit air to reenter the pile.

Spraying with water will lower the pile temperature, but experience has shown that the burning will normally reoccur once spraying ceases. This technique can be used in conjunction with blanketing. Once the pile temperature is lowered, the blanketing material is added, thus reducing the chances of the seal's cracking.

Grouting appears to be too expensive for the average coal operator to use in controlling a burning pile. This technique should be considered when property is threatened since it is able to isolate a fire, provide pile stability, and prevent the fire from spreading until it can be extinguished. Developing a suitable, easily applied surface seal appears to be the most economical procedure for extinguishing a burning pile.

SUMMARY

From the economic standpoint of the coal company, the most sensible approach is to prevent fires by using proper disposal methods and by quickly extinguishing hot spots before they spread. This will save the operator lost man hours, equipment, and materials. The cost of extinguishing a burning pile has been estimated between 50 cents and two dollars a ton, depending on various factors. This should be compared with the five to fifty cents a ton it costs to properly construct the refuse pile.

REFERENCE

Walton, John Wayne. "Coal Refuse Disposal—A Burning Problem," unpublished master's thesis, University of Cincinnati, 1969.

23.
EVALUATION OF ABATEMENT METHODS APPLICABLE TO COTTON GINS

Vernon P. Moore

Investigations Leader

and

Oliver L. McCaskill

Agricultural Engineer
Agricultural Engineering Research Division
Agricultural Research Service
U.S. Department of Agriculture
Leland, Mississippi

Only in recent years, ten or so, has the gin been considered by some as an air pollution problem. The sight of dust drifting from the gin and the distinctive odor of gin trash burning were at most considered to be necessary and welcome nuisances. It was symbolic of the harvest season which made it possible for the farm community to reap the benefits of a year's labor. Whether good or bad, this philosophy has changed. Instead of a gin being designed to the gin wall, it must now include a foreign matter collection system designed to meet state and federal regulations.

In the research program of the U.S. Department of Agriculture, an effort is made to anticipate the requirements and needs of the ginning industry. The need for waste collection equipment was no exception. A cooperative study was made with the Public Health Service in 1957 and published in 1960.[8] This study did little other than discuss the problem and give comparative emission rates for hand- and machine-picked cotton.

Since that time, hand harvesting has all but disappeared from the scene, and gin plants generally are of higher capacity. During the 1970-71 cotton season, machine picking accounted for 71% of the crop and machine stripping 26%. Hand picking and hand snapping accounted for only 3%. Although every bale of cotton will vary in trash content, laboratory tests have shown that the actual plant parts in a bale of seed cotton will be about 81 pounds for machine-picked cotton and 525 pounds for machine-stripped cotton.[9] Actually, the gin removes more than just dry plant parts—there are green bolls, short fibers, and dirt to be collected also. Therefore, the total actual amount of foreign matter removed per bale will be about 200 pounds for machine-picked and 840 pounds for machine-stripped cotton (Table 16). Since there is considerable variation within

Table 16. Weight of Material Removed during Ginning a 500-Pound Gross
Weight Bale from Cotton Harvested by Various Methods

Method of Harvest

	Hand Picked	Machine Picked	Machine Stripped	Hand Snapped
Percentage of crop*	2	71	26	1
Waste weight, pounds†	107	203	840	715

*Charges for Ginning Cotton, Costs of Selected Services Incident to
Marketing, and related Information, Season 1970–71, ERS and C&MS,
U.S.D.A.

†Calculated from seed cotton weights using seed-lint ratio furnished by
Cotton Breeders and corrected for 3% seed cotton moisture removal during
ginning. Weighted waste average for the crop: 322 pounds per bale.

harvesting methods and among years, the foreign matter range for each
harvesting method will be rather wide.

The composition of gin trash will undoubtedly vary with the type of
harvest, season, geographical location, and other factors. However, to get
some idea of the type of material being dealt with, trash from some late
season machine-picked Mississippi Delta cotton was analyzed.[10] It was
found that 96.49% of the trash was larger than 150 microns in size. Less
than 0.35% was smaller than 30 microns (Table 17).

Table 17. Gin Trash Composition of Late Season Machine-Picked Mississippi
Delta Cotton

Particle Size (microns)	Trash Content (percent)
>150	96.49
75–150	1.82
30– 75	1.34
10– 30	0.34
< 10	<0.01
Total	100.00

Almost without exception, the conveying systems in gins are pneumatic. Screw conveyors are used for seed in some cases and in and under machines for handling trash. They are the exception rather than the rule for conveying material outside the building. As a result, a large volume of air is required for the modern gin (Table 18). At a ginning rate of 12 to 15 bales per hour, a total of some 88,500 cfm is required. This volume is about equally divided between the high-pressure fans for handling seed cotton and trash and the low-pressure fans for handling fiber. Therefore, the problem at gins resolves itself to how to most economically collect 40 pounds of trash per minute from machine-picked cotton or 168 pounds from machine-stripped cotton when it is being conveyed by some 88,500 cubic feet of air in the case of a 12- to 15-bale-per-hour plant. Of course, some plants are capable of only six bales per hour while others will gin up to 30 or more.

Table 18. Typical Air Volume Requirements for Various Seed Cotton Trash-Handling Systems, Based on 12 to 15 Bales per Hour

System	Volume cfm
Seed Cotton and Trash Handling *	
Unloading system	8,500
No. 1 drying and cleaning system	9,000
No. 2 drying and cleaning system	9,000
Live overflow	4,000
Trash fan	3,000
Lint cleaner trash fan	10,000
Lint Handling†	
Lint cleaner exhaust	30,000
Condenser exhaust	15,000
Total	88,500

*High-pressure fan.
†Low-pressure fan.

DIMENSIONS OF A COMMERCIAL
TYPE, LARGE DIAMETER CYCLONE

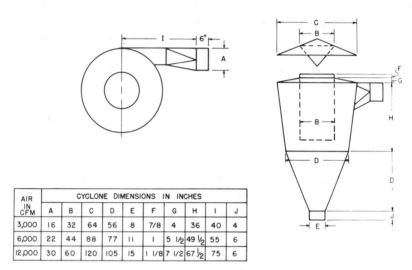

AIR IN CFM	CYCLONE DIMENSIONS IN INCHES									
	A	B	C	D	E	F	G	H	I	J
3,000	16	32	64	56	8	7/8	4	36	40	4
6,000	22	44	88	77	11	1	5 1/2	49 1/2	55	6
12,000	30	60	120	105	15	1 1/8	7 1/2	67 1/2	75	6

Figure 75. Dimensions of a commercial large-diameter cyclone

RELATIVE DIMENSIONS FOR A
SMALL DIAMETER DESIGN CYCLONE

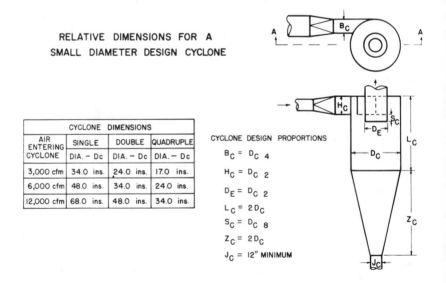

CYCLONE DIMENSIONS			
AIR ENTERING CYCLONE	SINGLE DIA. − Dc	DOUBLE DIA. − Dc	QUADRUPLE DIA. − Dc
3,000 cfm	34.0 ins.	24.0 ins.	17.0 ins.
6,000 cfm	48.0 ins.	34.0 ins.	24.0 ins.
12,000 cfm	68.0 ins.	48.0 ins.	34.0 ins.

CYCLONE DESIGN PROPORTIONS

$$B_C = \frac{D_C}{4}$$

$$H_C = \frac{D_C}{2}$$

$$D_E = \frac{D_C}{2}$$

$$L_C = 2 D_C$$

$$S_C = \frac{D_C}{8}$$

$$Z_C = 2 D_C$$

$$J_C = 12'' \text{ MINIMUM}$$

Figure 76. Atomic-energy-designed high-efficiency cyclone

The cyclone was the first, and for some time the only, type collector used at gins. It was what has become known as the "large-diameter" design which has been widely used for many years for the collection of burs and other heavy plant parts (Figure 75).

During the early 1950s, the AEC or "small-diameter" high-efficiency cyclone was introduced to the ginning industry (Figure 76). Like the conventional cyclone, it is used only on high-pressure fans since it operates more efficiently at ranges from about 3 and 5 inches static pressure (Figure 77). In recent years, this cyclone has become the standard of the industry, and it is almost without exception the only type being installed.[4]

It was an accepted fact that the small-diameter cyclone was more efficient than the large-diameter model, but with the increased emphasis on pollution abatement, it was found desirable to determine just how efficient it was on gin trash. Various state air control boards began to request detailed information as they developed regulations. As a result, cyclone evaluation studies were made by the Ginning Laboratories.

Tests at the Ginning Laboratory at Mesilla Park, New Mexico, on machine-picked and -stripped cottons showed that trash input rates, inlet air velocities, and the size distribution of the trash all influenced the operation of a cyclone.[3] There was little difference in collection efficiency between a trash exit discharging into an airtight container and an exit discharging into an open container. Two sizes of cyclone trash exits were tested to determine the influence of exit size on collection efficiency. A 7½-inch diameter and a 12-inch diameter exit were tested on a 30-inch diameter cyclone. There was no difference in collection efficiency between the two sizes. A 12-inch minimum diameter outlet is now being recommended as standard for gin installations because of the less likelihood of sticks bridging the opening and causing chokeage.

The tests showed that the small-diameter cyclone will maintain an overall collection efficiency of 99.9% or greater on gin trash over a wide range of operation conditions. An inlet vane, inlet helix, and skimmer were tested to determine their effect on efficiency. The inlet vane and helix reduced pressure drop and efficiency while the skimmer had no effect.

A three-year study was carried out at the Ginning Research Laboratory at Stoneville, Mississippi, on the AEC cyclone using machine-picked cotton.[10] These tests showed that the design was virtually 100% efficient on particles larger than 20 microns in diameter. Particles smaller than 20 microns were partly collected in decreasing amounts as they became smaller (Figure 78). In these studies, the overall collection efficiency on small gin trash was 99.927%, which is about the same as was indicated by earlier studies.

These tests may be summarized as follows.

1. *Operating air velocities significantly affect the cyclone's collection efficiency* with greatest efficiencies being in the range of 3,000 feet per minute.

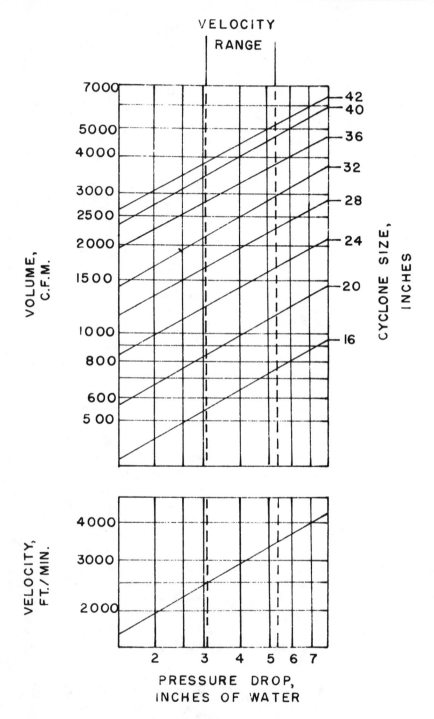

Figure 77. Pressure drop for atomic-energy-designed cyclone

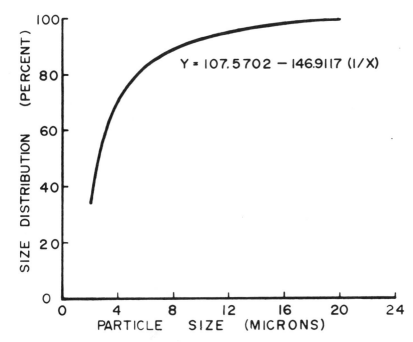

Figure 78. Particle size distribution from atomic-energy-designed cyclones

2. *For the high input concentrations*, neither trash size nor input concentrations significantly affect the collection efficiency of the cyclone.

3. *For input concentration normally encountered by the cyclone at a cotton gin* (less than 105 grains per cubic foot), the experiments indicated that trash sizes, operating air velocities, and input concentrations had a significant effect on the collection efficiency of the cyclone.

4. *For the high input concentrations*, test results revealed that operating air velocities and input concentrations significantly affected the dust concentration in the exhaust air from the cyclone.

5. *For the low input concentrations*, the trash size had no significant effect on dust concentration in the exhaust air from the cyclone.

6. *For the low input concentrations* (less than 105 grains per cubic foot), data revealed that trash sizes, operating air velocities, and the input concentrations significantly affected the dust concentration in the exhaust air from the cyclone.

The low-pressure fans used in the lint-handling systems, depending on type, will safely operate against a system static pressure of from one to five inches w.g. Therefore, it is not possible to use cyclone collectors with them. Whereas all of the large foreign matter in the gin plant is handled by high-pressure fans, the foreign matter carried by the exhaust air from these fans is primarily lint, small pieces of vegetative matter, and soil particles or "dust."

These large quantities of dust- and lint-laden air create a nuisance as well as a fire hazard around the plant. The lint will blanket the ground, trees, and other structures near the exhausts. Not only is it unsightly, it will burn rapidly if ignited. Screen cages were developed for use in the low-humidity areas of the belt to collect the "lint fly."[4] Their collection efficiency was quite low and attempts to adopt them for use under high-humidity conditions were generally unsuccessful.

Alberson and Baker developed the inline filter, versions of which, in the case of new installations, have all but replaced the screen cage.[1] The device consists of a stationary filter screen made of 105-mesh stainless steel bolting cloth (Figure 79). As material builds up on the filter, the increase in static pressure actuates the wiper-brush motor. The brush wipes the material from the screen into a small negative-pressure air line, which carries it to a cyclone.

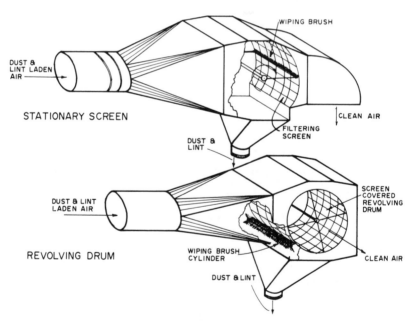

Figure 79. Schematic of a fixed-screen and revolving-drum inline filter

In addition to the original stationary screen type, there are two other versions of the same principle on the market at the present time.[2] One has a revolving screen that is built similar to a conventional condenser (Figure

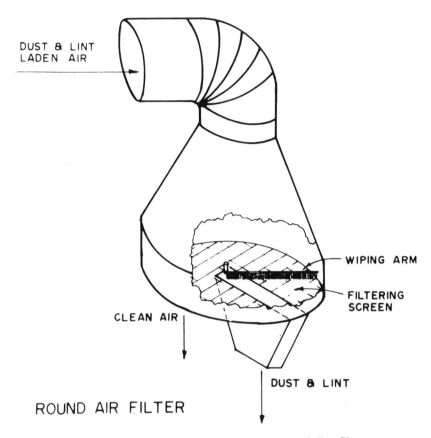

Figure 80. Schematic of a flat, round-screen, inline filter

80). The second type has a flat, round screen with the wiper-arm motor mounted in the center of the screen (Figure 81). In comparative tests, all three types had a collection efficiency of 99% on leaf fly and an overall collection efficiency of about 81% using stripped cotton. The original designs used 105-mesh bolting cloth having an open area of 46.9%. Many of the units now being manufactured use 70-mesh stainless steel bolting cloth having an open area of 54.9%. Some manufacturers use pressure switches to actuate the wiper arm or brush; others omit the pressure switch, allowing the doffing mechanism to operate continuously. When doffing is continuous, the filter efficiency is greatly reduced and more dust is emitted to the atmosphere.[6] Therefore, pressure switches are recommended although there is a slight added cost and some preventative maintenance is required.

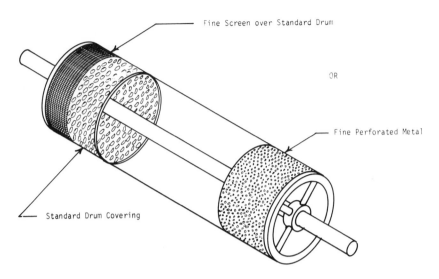

Figure 81. Condenser drum covered with fine screen or fine perforated metal

When a gin is located where the dust does not pose a major problem, some ginners are covering their condenser drums with 70-mesh bolting cloth over the conventional covering to stop the lint fly (Figure 81). This reduced the lint fly and dust concentration in the exhaust air from 50.8 to 21.6 grains per 1,000 cubic feet in tests made at Stoneville, Mississippi.[5] By covering the condenser with perforated metal having .033-inch diameter openings and a 20% open area in lieu of the conventional covering material, the dust concentration is reduced from 50.8 to 32.0 grains per 1,000 cubic feet of exhaust air. One of the gin machinery manufacturers has changed the specifications for its condenser coverings, and now the 70-mesh bolting cloth is optional on large condensers and the small opening perforated metal is standard on its lint-cleaner condensers.

Tests have been run on both machine-picked[6] and stripped cotton[7] to determine rate of emission from a gin plant. A recommended amount of gin machinery was used for the type of cotton being processed. The trash collection system included conventional AEC high efficiency cyclones on the high pressure fan exhausts. The discharge pipes from the cyclone hoods were sized to give a velocity within the range of the high volume air sampler for isokinetic sampling. Both sampling and filter analysis were in accordance with accepted procedure. The emission rate for machine-picked cotton was calculated to 13.5 pounds per hour at a 10-bale-per-hour ginning rate (Table 19). The highest concentration of material is from the unloading fan and lint-cleaner condenser exhausts. The same emission pattern held true for stripped cotton although the rate was substantially higher, averaging about 15 and 23 pounds respectively for early and late season cotton. The emission rate range was quite wide on all of the studies. However, using the data from these studies and from the collection efficiencies of the various types of abatement equipment, some idea can be obtained of what is necessary for a gin to meet specific state regulations.

Table 19. Total Emission Adjusted to a 10-Bale-per-Hour Plant Ginning Machine-Picked Cotton

Exhausts	Total Emission	
	High* Lbs/hr	Average Lbs/hr
Cyclone Exhausts		
Unloader fan	5.41	1.14
6-cylinder cleaner		
and stick machine	.14	.06
6-cylinder cleaner	.08	.04
Trash fan	.16	.08
Condenser Exhausts		
3 No. 1 lint cleaners	13.92	6.98
3 No. 2 lint cleaners	5.62	2.08
Battery	2.10	1.33
Lint cleaner waste	2.55	1.80
Total	29.98	13.51

*Highest single emission of all tests.

Acknowledgement

This chapter was originally developed for the U.S. Department of Agriculture and is published herein with the permission of the U.S.D.A.

REFERENCES

1. Alberson, David M., and Roy V. Baker. "An Inline Filter for Collecting Cotton Gin Condenser Air Pollutants." Washington, D.C.: U.S. Dept. Agr., Agr. Res. Serv., ARS 42-103, 1964. 16 pp.
2. Baker, Roy V., and Calvin B. Parnell, Jr. "Three Types of Condenser Exhaust Filters for Fly Lint and Dust Control at Cotton Gins." Washington, D.C.: U.S. Dept. Agr., Agr. Res. Serv., ARS 42-192, 1971. 13 pp.

3. Baker, Roy V., and V. L. Stedronsky. "Gin Trash Collection Efficiency of Small-Diameter Cyclones." Washington, D.C.: U.S. Dept. Agr., Agr. Res. Serv., ARS 42-133, 1967. 16 pp.

4. Harrell, E. A, and V. P. Moore. "Trash Collecting Systems for Cotton Gins." Washington, D.C.: U.S. Dept. Agr., Agr. Res. Serv., ARS 42-62, 1962. 22 pp.

5. MaCaskill, Oliver L., and Vernon P. Moore. "Elimination of Lint Fly—A Progress Report," *The Cotton Gin and Oil Mill Press*, December 21, 1966.

6. McCaskill, Oliver L., and Richard A. Wesley. "Tests Conducted on Exhausts of Gins Handling Machine-Picked Cotton," *The Cotton Gin and Oil Mill Press*, September 5, 1970.

7. Parnell, Calvin B., Jr., and Roy V. Baker. "Particulate Matter Emissions by a Cotton Gin: A Study," *The Cotton Gin and Oil Mill Press*, April 17, 1971.

8. United States Department of Health, Education and Welfare. "Airborne Particulate Emissions from Cotton Ginning Operations." Cincinnati, Ohio: U.S. Pub. Health Serv. Tech. Report A 60-5, 1970.

9. United States Department of Health, Education and Welfare. "Control and Disposal of Cotton Ginning Wastes." Cincinnati, Ohio: U.S. Pub. Health Serv., Bur. of Disease Prevention and Environmental Control, 1967.

10. Wesley, Richard A., Wm. D. Mayfield, and Oliver L. McCaskill. "An Evaluation of the Cyclone Collector for Cotton Gins." Washington, D.C.: U.S. Dept. Agr., Agr. Res. Serv., Tech. Bul. 1439, 1972. 13 pp.

Section V

Chemical and Wood Products Industry

24.
CONTROL OF ACRYLIC ODORS FROM PLASTIC MANUFACTURING

Leroy G. Fox

Chemical Engineer
Rohm and Haas Tennessee Incorporated
Knoxville, Tennessee

The emission of odors from industries is not a rare problem. We shall describe here the nature of a specific problem arising out of acrylic plastic manufacturing at the Knoxville, Tennessee, plant of Rohm and Haas Tennessee Incorporated and narrate the strategy used to approach a satisfactory degree of control.

The subjective nature of odors gives rise to many difficulties in their quantitative measurement, their meaningful control by regulation, and implementation of an adequate control strategy for them.

Knox County, Tennessee, contains in its Air Pollution Control Regulations of 1969 the restriction of "objectionable" odors beyond the property line from which such emissions occur and which are in sufficient quantity and of such characteristics and duration as to be, or to tend to be, injurious to the public welfare, to the health of human, plant, or animal life, or to property, or which unreasonably interfere with the enjoyment of life and property. The regulation further states, "An odor shall be deemed objectionable when 30% or more of the people exposed to it believe it to be objectionable in their usual places of occupancy based on a sample size of at least 20 people, or if fewer than 20 people are exposed, 75% must believe it to be objectionable, or when air occurring beyond the property line contains such odorous matter as may be detectable when it is diluted with 8 or more volumes of odor-free air." As we see, the regulation of odor by law has been partly quantified, but the word "objectionable" in the regulation still recognizes the diverse responses of human beings to sensing an odor.

An attempt can be made to quantify and compare the intensity of an odor. For instance, one cubic foot of ammonia gas mixed with several hundred cubic feet of air would still have a pungent odor, but if one cubic foot of ammonia were diluted with several million cubic feet of odor-free air, the mixture would no longer have any detectable odor. The dilution which must be carried out in order to render an odor just barely perceptible to the human nose, as measured by a panel of experienced people, is known as the odor threshold. These threshold values are usually expressed as parts per million by volume. The odor threshold of two of our acrylic monomers used in the largest volume at the Knoxville plant are ethyl acrylate, 0.5 part per billion, and methyl methacrylate, 0.21 part per million. These values could also be expressed as one part in two billion for ethyl acrylate, and one part in five million for methyl methacrylate odors to be detectable by an "average" human nose—if such a thing exists. These odor threshold values are at concentrations much below the levels affecting health. The maximum atmospheric concentration for an 8-hour working day without injury to health is for ethyl acrylate, 25 parts per million, and for methyl methacrylate, 100 parts per million. Because of their irritating effects at these concentrations, these materials act as their own warning agents.

The potential at the Knoxville plant for release of acrylic odors is immense. Over ten million pounds of acrylic and related liquid monomers such as methacrylates are processed every month, with most of these materials being handled more than one time; that is, they are transported from one vessel to another. In the conversion of the monomers into long chain polymer molecules the odor is lost; the polymers are odorless solids. These useful raw materials are used in the manufacture of acrylic plastics such as Plexiglas acrylic sheet, and Rhoplex dispersions of acrylic polymers in water. The popular and useful water-based paints for both indoor and outdoor application owe their existence to Rohm and Haas pioneering in this field.

Now, let us look at the odor problem in its environmental setting. The 32-acre Knoxville plant is located one mile west of Knoxville's City Hall. Within a radius of 0.5 mile from the plant there are major U.S. highways and industrial, commercial, and residential properties. Low ridges running approximately northeast to southwest lie on both sides of the shallow valley in which the plant is situated. These ridges are slightly higher than the rooftops of the plant. This topography, in addition to the tendency of the east Tennessee area to suffer from general low wind velocities and frequent stagnations in the atmosphere throughout the year, made us skeptical of relying on diffusion alone to dilute odors to a level at which they would not be detected most of the time. One must be cognizant of the environmental background to gain an understanding of the considerations which must go into an odor management program.

The first step in reducing the emissions of odor was to find out exactly how much of the acrylic monomers were being emitted to the air. We conducted an emission inventory in 1968 and 1969, using the gas laws as a basis for estimating the loss of these materials depending on the temperature and pressure under which they were handled or processed and the amount of the vapors which were displaced or which escaped into the atmosphere. This was a relatively easy step, and it gave us a good insight into the odor potential from various areas of operation. It was also instructive in setting priorities for management of the larger sources as they were defined.

At about the same time, we began to conduct objective and subjective tests for the presence of odors beyond the plant property lines. A monitoring network was set up with eight sampling points located 45° apart on circles initially of 1.0 mile radius and of 0.5 mile radius from the plant, with a later addition of sampling team of two persons visited eight points on the circles at random times during each week. At each sample point, they noted the presence or absence of odors, and they also took a grab sample of the ambient air at that point in a 50-ml glass syringe. The syringes were promptly returned to the plant and examined by gas chromotography in our laboratories for the presence of acrylic monomers. In only a few cases were these monomers detected downwind from the plant, and then only 0.25 mile away. Curiously, the sampling team did not invariably note acrylic odors being present at the time that positive syringe samples were obtained. Beginning in 1972, a "Telematic" portable sampling pump and a bubbler were used to obtain samples over a much longer period than represented by the grab samples. Such bubblers have the effect of depressing peak values, but they totalize emissions absorbed over a period of several hours. Concentrations we believe to be as low as 0.1 ppm methyl methacrylate are detectable by the syringe method. The maximum concentration of methyl methacrylate actually detected was about 1 ppm. In addition to the acrylic monomers detected occasionally, there was almost always a considerable background of other materials believed to be mostly hydrocarbons from automotive emissions. Once we even detected a leak in a gas main due to encountering a high concentration of methane above the normal background. Field use of the "Scentometer," a commercial instrument capable of performing manifold dilutions with odor-free air, was of only limited value to us because the odors never persisted long enough to make the specified series of dilutions within the instrument.

We also began in 1969 to accumulate data on the odor complaint pattern in both time and space. We installed instruments to provide a continuous record of wind direction and velocity. Each complaint was investigated and the position of the observation was located on a wall map. The complaints were generally located downwind and within a 1.0 mile radius of the plant. Two unexplained reports of acrylic odors were received from downwind points as far as 2.5 miles from the plant. A few anomalous complaints were received which were distant from the plant and upwind on a day in which the wind was strong and steady. Perhaps someone nearby had been burning plastic scraps or had been using power tools on acrylic plastics. An analysis of the complaints has been very helpful in providing a background of exactly what the community is detecting and where and how often. These findings will serve as a bench mark to measure progress

in our odor abatement activities. Even complaints, if evaluated construc-
tively, are a real aid in an odor management program.

Techniques similar to the syringe method used in the monitoring network enabled us to determine the quantitative concentration of acrylic
monomers in the various vents and ventilation discharges from the plant
operating areas and from our storage tanks. Some of the larger emissions
which we measured were fed into the Bosanquet-Pearson diffusion equations to check the observations which were being made on the area
surrounding the plant. We used the results obtained to evaluate possible
strategies of odor control, but we were very cautious in applying any of
our results in an odor management program because of the topographical
features of the plant neighborhood, noted previously.

As we accumulated data from 1969 through 1971, we were simultaneously introducing process changes to reduce or eliminate odor emissions.
Because of the unique character of the acrylate monomers in producing
durable plastics for outdoor service, it was not practical to try a most
obvious technique: alternate raw materials. We did make changes in some
of the equipment, such as providing a return loop for vapor displaced from
a storage tank as the contents of a tank car were pumped into the tank.
The displaced vapors were thereby returned to the tank car, resulting in a
great reduction in net emissions from such operations. Changes in processing techniques in some chemical processing equipment were implemented.
Where the nature of the process permitted, the vessels containing acrylic
monomers were pressurized, operated with minimum opening of manholes
for access, and with minimum use of gases to purge the reaction vessels.
Because of certain features of the processes, it was not possible to conduct
the operations under a completely tight system. We also increased our
vigilance to prevent unwarranted discharges from pressure relief valves and
breakdowns of piping, pumps, and meters with resultant leaks of odorous
materials.

As we began to see the limitations facing us on process changes to
control odors and as we acquired more knowledge of exactly where the
odors were coming from, we were able to group the techniques for coping
with them. We considered all of the odor control methods which we will
list here, although in most cases we did not make a comprehensive cost
study of each.

A fundamental strategy was to maintain as high a concentration of the
odor emissions prior to treatment as we could. The reasons for this were
that a minimum volume of contaminated air would have to be moved; if
incineration was used, there would be a minimum amount of fuel needed
to heat the contaminated air to the combustion point; and absorbing or
adsorbing equipment would be of minimum dimension and cost. One can
not afford to let the odors escape into a large area where their pursuit
would result in handling an excessive volume of relatively low concentrations of odor.

The use of sorption processes to abate acrylic odors looked particularly attractive because of their adaptability to wide ranges of concentrations and fairly wide range of tolerance to gas velocities. We looked at adsorption methods, particularly the use of activated carbon, which had been shown in early experimentation to be highly efficient. An inherent disadvantage with the use of activated carbon is its high heat of adsorption, which could lead to possible ignition of flammable vapors. In the case of the reactive acrylic monomers, this heat could promote conversion of the adsorbed acrylates to a solid polymer, thus rendering the adsorption bed inactive and irrecoverable.

Our scouting studies indicated that the nonregenerative adsorbers using activated carbon may be practical for small, isolated sources with low concentrations of acrylates, but we felt that such use would lead to maintenance problems for the very small units we visualized, with a tendency to neglect them over a long period. The regenerative adsorbers seemed to have an advantage of somewhat lower operating cost but higher capital costs than the nonregenerative type. The regenerative adsorbers give rise to the problem of disposal of the adsorbed acrylates, assuming they had not polymerized, after the activated carbon had been regenerated; that is, the process does not destroy any acrylates but simply takes the materials from the air and concentrates them on the solid carbon. The depleted carbon could be sent, of course, to the supplier of the carbon for regeneration. Because of the tendency of acrylic monomers to polymerize, adsorption with carbon did not appear to be particularly attractive, and so we looked to other methods.

We considered various absorbing processes for acrylic odors. One of our sister plants had installed an efficient absorber for ethyl acrylate suitable for such applications as tank vents, using as absorbing liquid the ethyl acrylate monomer itself, chilled to approximately $0°$ F where its vapor pressure is about 3 mm of mercury compared with 50 mm vapor pressure at $86°$ F. Such a refrigerated scrubber, a form of self-absorption, resulted in fairly high efficiencies of about 95%, but its practical application was limited to the condensation of pure monomers because the scrubbing liquid had approximately the same composition as the material stored in the tank. The method has the advantage of not generating directly any additional wastes. Many of our processes generate emissions of mixed and variable compositions which are not suitable for recycle to any process.

Absorption of acrylic odors by a nonreactive liquid solvent, which would be regenerated by a suitable process in an accessory to the absorber, was given an appraisal. Regeneration costs of possibly 25 cents per pound of monomer absorbed and the finding of a means of disposal of the absorbate separated from the solvent made us skeptical of the practicability of this route.

The acrylic monomers which we handle in large volume are reactive with aqueous sodium hydroxide solution even at room temperature. The reaction product of the caustic soda with the acrylates is odorless. The acrylates are also reactive, for example, with potassium permanagnate and with chlorine: the permanganate seemed unattractive because of the ultimate discharge of manganese compounds to the plant sewers to be treated by the city treatment plant, and chlorine handling would have introduced hazards and costs which we were not willing to accept. Our research laboratories and one sister plant had conducted some early experiments with a caustic soda absorbing tower with efficiencies of over 95% removal of acrylates. Because of the relative economy of using caustic soda, which is a common chemical at our plant, of its relatively innocuous effects when the neutralized, depleted solution is discharged to the sewers, and of the simplicity and flexibility of the method, we designed and built a prototype scrubber with dimensions of 10 inches in diameter and 10 feet high, packed with 1-inch Raschig rings and using caustic soda to react with ethyl acrylate odors. This prototype scrubber has been in operation almost a year, completely removing detectable ethyl acrylate odors from the vent of a large storage tank. Measured by samples taken on the input and exit sides, the scrubbing tower has demonstrated an efficiency of 100% or very close to 100% removal when operating at approximately 60° F with concentrations of caustic soda solution ranging from 2.5 to 10%. Depletion of this type of scrubbing liquid unfortunately also occurs due to its reactivity with carbon dioxide in the air. A significant amount of consumption of the caustic soda would be expected from this by-product effect. We determined that the 5-day biological oxygen demand (BOD_5) of the absorbing solution was very close to a 1:1 ratio with the weight of ethyl acrylate absorbed; that is, one pound of BOD_5 was generated for each pound of ethyl acrylate absorbed. Because this route seemed particularly effective, we continued to operate the prototype scrubber in this service while we examined other alternate means of controlling the odors.

The surest way to completely kill acrylic or other industrial odors is to destroy the molecule. Probably the most effective way to destroy the acrylic and other organic odors is by combustion processes. Because we handle large volumes of highly flammable liquids at our Knoxville plant, combustion processes are always looked upon with suspicion. From the broad view, the connection of any maintained fire or source of ignition, which could propagate flame to storage tanks or processing vessels, seemed rather scary. We considered a catalytic combustion process but felt that the high fuel cost in money and in consumption of the natural gas resources required to maintain the catalyst at peak operating temperature would be prohibitive because the fuel value contributed by the very low concentrations of acrylic monomers would be negligible. Use of a flare stack as the terminus for all collected odor emissions, with an open flame at the top of a tall tower, seemed decidedly unattractive because of the highly visible and ever-present flame and the extra air pollution caused by its combustion products. Our initial look at combustion of the low concentrations of acrylic monomers in a furnace or combustion chamber disclosed the same drawbacks as we had encountered in considering the use of catalytic combustion. As we went back over the alternates available to us for efficient control of odors at the lowest possible cost, the combustion processes seemed more and more attractive.

Two other methods for odor control, chiefly for acute emissions from a spill, were considered. Masking or neutralizing agents for acrylic odors have been claimed by various suppliers; however, we felt that these were quite expensive, and they added additional contaminants to the air with no assurance that the masking agents themselves would not be objectionable to some of our neighbors. We have found that certain types of foam fire extinguishers will provide a stable blanket over a spill of some of the acrylic monomers, but such a technique is only an expedient at best.

The final selection of control equipment for odor at the Knoxville plant was made after considerable discussion with our Corporate Engineering Division and our sister plants which have emission potentials for acrylic odors similar to those at Knoxville. We abandoned plans to use a refrigerated liquid scrubber on the basis of projected operating and maintenance costs coupled with an efficiency as low as 95%. We have a demonstrated performer in the prototype caustic soda scrubber which we will continue to operate to control emissions from a cluster of storage tanks.

A study of the volume of plantwide emissions of acrylic odors compared with the volume of air used in combustion of natural gas or fuel oil in our steam boilers indicated that the concentrated emissions of odor from the entire plant area could be incinerated with negligible operating cost by feeding them into the combustion chambers of our three large steam boilers. Processing of various kinds carried out in our plant is contingent upon having steam available to supply heat and power, and the boilers operate continuously throughout the year. At any time when the steam boilers are shut down, our plant is shut down. Since the various areas in the plant have slightly different compositions of vapors and odors emitted from the processes, a study of each separate area was made to see if its output could be incorporated into the combustion air. All of these projections were favorable to the method, and we suggested the use of this type of combustion operation to our corporate executives. We received their approval of this technique, which is essentially no different from the package incineration of any other flammable gas or vapor, and the same barriers and controls to prevent flame propagation and for detection and control of flammable mixtures are required. The various conduits containing flame arrestors, to convey the collected acrylic odors to the boilers are being engineered and designed. The conduits will be up to 400 feet in length and from 18 to 36 inches in diameter. The processes having odor emissions are being connected into a series of collection manifolds. The odor collection system, of course, is independent of the actual abatement equipment, and the collected vapors could just as easily have been conducted to a caustic soda scrubbing device as to the boilers.

This review has touched on only the high spots of the control program. We believe that a similar program could be applied to many industrial odor problems. Good maintenance and management of the equipment being installed this year will be required to avoid any significant acrylic odor emissions from the Knoxville plant.

25.
DEFINING AND CONTROLLING HYDROCARBONS FROM PROCESS EMISSIONS

William R. King

Chief, Chemical Section, Division of Applied Technology
Environmental Protection Agency, Office of Air Programs
Research Triangle Park, North Carolina

This paper will cover two specific topics in detail: (1) EPA's current work in the areas of hydrocarbon-problem definition and control-method development, and (2) the economics of solvent substitution versus recovery for relatively large-volume solvent degreasers. First, a quick review of the importance of stationary sources in the hydrocarbon emissions picture is in order (Figure 82).[1] As you can see, mobile sources were the major hydrocarbon emitters in 1968. Open burning (controlled and uncontrolled), industrial processes, solvent evaporation, solid waste disposal, and gasoline marketing together contributed the remaining 50% of the emissions.

The Clean Air Amendments of 1970 require the automobile industry to reduce emissions drastically within the framework of a relatively inflexible time table. EPA estimates that these rules will result in the reduction of automobile emissions by 1980 to 3.6×10^6 tons per year. Since total emissions from all sources have been estimated at 25×10^6 tons per year for 1980, nonmobile sources will then become the major hydrocarbon emitters.

Traditionally, hydrocarbons have been one of the "Big 5" pollutant categories that the Federal air pollution program has worried about. The other four are sulfur oxides, nitrogen oxides, carbon monoxide, and particulates. Frankly, most of the previous work has been concerned with mobile sources because of their overriding contribution to the pollution problem. In 1970, however, EPA began a serious attack on stationary-source problems.

EPA has contracted with MSA Research Corporation (MSAR), a division of Mine Safety Appliances Company, to perform a "Hydrocarbon Pollutant Systems Study."[5] The purpose of this study is to delineate the problem of stationary-source hydrocarbon emissions and to develop a problem-solving program for the control of hydrocarbon pollutants from those sources.

To achieve the objectives, a systems study has been developed to integrate the results of several intermediate tasks into a model. The model

251

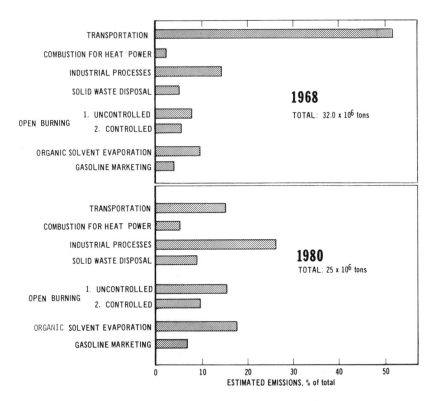

Figure 82. Sources of Hydrocarbon Emissions, 1968 and 1980

output will be an overall program leading, if necessary, to projects to demonstrate control hardware. The program's intermediate tasks are:

1. To identify, characterize, and rank all significant stationary sources of hydrocarbon emissions. Sources and emissions will be ranked both according to quantity and adverse effects such as health, physiological, and vegetative damage, and according to their contribution to photochemical smog.

2. To characterize the effluent streams from major stationary sources in terms of concentrations, flow rates, temperature, and other parameters that could influence control techniques.

3. To evaluate, both technically and economically, the existing and potential control technology for selected, major emission sources.

The majority of the contractor's effort, about 75%, will be expended on source definition.

The input for these intermediate phases of the study is being obtained from a critical review of the open literature and surveys of both the public

and private sectors. The latter are being accomplished by sending questionnaires to state, regional, and municipal pollution-control agencies and to
selected trade and industrial associations.

MSAR's studies of the adverse effects of organic pollutants have been
concentrated in three areas: photochemical-smog contribution, vegetative
damage, and effects on human health.

The contribution of certain hydrocarbon compounds to the photochemical reactions leading to smog formation has been recognized since the
late 1940s. A large number of studies have been made to elucidate the
various reactions and the relative reactivities of specific organic compounds.
These studies, conducted primarily in smog chambers, have been reviewed
by MSAR, and the result is a list ranking the possible smog-forming organic
compounds, classifying them either by specific name or by chemical group.
Lack of agreement among the various investigators regarding the proper
smog-chamber test conditions and the relative weights to be applied to each
of the indices of reactivity (eye irritation response, NO_2-production rates,
etc.) tend to complicate the ranking of specific compounds. However, a
consensus seems to exist on the ranking of chemical groups.

Reports of vegetative damage by organic compounds have generally
indicated only ethylene and the oxidants formed from photochemical
reactions.

The adverse health effects of organics are being studied by the Industrial
Health Foundation, a study subcontractor, who is making use of the available documentation on Threshold Limit Values (TLVs) published by the
American Conference of Governmental and Industrial Hygienists and com-
plementing these data with animal-toxicity studies, eye-irritation indices,
and lethal-dosage data. The TLVs are recommendations for 8-hour industrial exposure; hence other data and expert medical judgment are necessary
to extrapolate those short-term data to continuous low-level exposure
values.

In order to obtain possibly unpublished information on organic emissions and control procedures, MSAR decided to use two simple question-
naire surveys. The first was directed to state, regional, and municipal
pollution-control agencies, the second to selected industry and trade
associations.

Both questionnaires were designed to determine (1) the type and availability of any emission surveys or inventory studies, either completed or in
progress; (2) what specific problem areas are believed to be related to
organic emissions; (3) the type and extent of existing controls; and (4) the
amounts of solid waste disposed of by combustion methods.

The responses to the questionnaires have been fairly good. More than
50% of the questionnaires were returned from the control agencies and
30+% from the trade associations. The information obtained, however, has

been somewhat disappointing; about 50% of the control-agency returns and 75% of the trade-association returns indicate essentially no available data.

MSAR's estimates of total emissions from fuel combustion (coal, fuel oil, and natural gas) agree reasonably well, except for estimates of natural-gas emissions, with estimates in EPA's hydrocarbon control techniques document. Recent data indicate significant organic emissions from natural-gas combustion, whereas EPA's estimates show essentially none. Problems arise when one attempts to break these total emissions into component groups such as simple hydrocarbons, aldehydes, ketones, and organic acids because stack-sampling data do not agree. More definitive sampling studies of the various organic components of flue gases from actual combustion sources are needed in order to clarify those discrepancies.

Industrial processes are another major category of sources that may have significantly higher emissions than previously estimated.[1] EPA's earlier estimates were based on petroleum refining, petrochemical processing, natural-gas-liquids-recovery, and coke manufacturing. Hard data on industrial process losses are few and far between; however, some information on previously ignored sources has been uncovered. In the wood-pulping industry, for example, a multitude of publications are available on the emission of inorganic gases and reduced-sulfur compounds from pulping operations, but only two papers have been found which mention evaporative losses of organics such as turpentine, methanol, acetone, and other unidentified compounds. Turpentine losses from incomplete recovery operations may run as high as 50,000 tons per year.[3] Methanol emissions from a modern 500-ton-per-day Kraft mill exceed 0.5 ton per day.[4]

EPA is still in the problem-definition stage as far as stationary-source hydrocarbon emissions are concerned. We have begun a hardware-oriented study, however, without waiting to identify and characterize all hydrocarbon pollutants; its title is *Package Sorption Device System Study*.[6] The objectives are:

1. To determine the nationwide extent and type of gaseous air pollution from small, fixed sources.
2. To evaluate the existing technology of small, fixed-bed, sorbent- and catalytic-combustion-package devices for the control of pollution from these sources.
3. To assess the potential of new technology.
4. To recommend a program of study that will advance package-sorption and catalytic-control technology.
5. To prepare a handbook presenting the collected knowledge on air pollution from small sources and on package-control technology.

Note that the objectives do not limit the study to hydrocarbons, but, realistically, most of the applications of these devices are to hydrocarbon-containing streams.

POLLUTION CONTROL   255

Table 20. Industries of Interest to Package Sorption Device System Study

Industry

Surface coating	Fish rendering
Degreasing	Canning
Dry cleaning	Coffee roasting
Graphic arts	Restaurants
Plastics	Tobacco curing
Rubber products	Baking
Extraction of vegetable oils	Tanning
Adhesives	Smoke houses
Paper coatings	Breweries and distilleries
Automobile repair and paint shops	Incinerators
Animal rendering	Sulfite paper mills

The industries listed in Table 20 are, at least in part, within the scope of the program. Some plants within these industries are definitely too large to be considered. The exact boundary between too-large and within-scope is somewhat arbitrary. In this study, the upper limit was set to include plants that do not exhaust over 10,000 cubic feet of pure air per minute.

The first ten industries listed in Table 20 are solvent users and appear to emit the largest amounts of pollutants into the atmosphere, although not necessarily the most obnoxious or toxic pollutants. The next ten industries do not appear to emit as much material but some of them, such as the rendering industries, emit obnoxious odors, and hence are under great pressure to clean their exhaust air streams. The air pollution problems of these middle ten are expected to be more difficult to correct because of the large amounts of moisture in their exhaust air streams, low pollutant concentrations, and the necessity for reducing odor-producing effluents to very low levels. The pollutants from the last two industries seem to be predominantly—although not entirely—inorganic.

The existing technology review has studied all types of sorbents, *i.e.,* silica gel, alumina gel, molecular sieves, chemisorbents, activated carbons, and impregnated activated carbons. The sorbents other than activated carbon are applicable only under special circumstances. The polar adsorbents, silica gel and alumina gel, are effective in moist atmospheres only for strong acids. Chemisorbents and impregnated carbon cannot normally be regenerated. Thus far, the pollution-emission survey has not revealed any special circumstances where other adsorbents would be more suitable than plain, activated carbons; hence, the efforts have been concentrated on the study of activated-carbon systems.

According to MSAR, the main problem with activated-carbon systems is the cost of regenerating them. Economic analyses already conducted show

that saturated-steam regeneration which is effective for solvent-recovery
operations at concentrations above 1,000 parts per million (ppm) becomes
costly at concentrations in the range 0.1 to 500 ppm. The latter is the
concentration range of greatest concern in control of air pollution from the
industries under consideration. Current studies based on physical chemistry
and engineering principles are aimed at determining the most economical
methods for accomplishing the adsorption-regeneration cycles.

Promulgation of Los Angeles' famous Rule 66 and the inclusion of
trichloroethylene on the list of photochemically active compounds raised
considerable worry—perhaps panic would be a better word—in the ranks of
the trichloroethylene manufacturers. They thought that most of the sol-
vent-degreaser users would choose substitution rather than recovery. I
understand that in Los Angeles this has happened, but it is not obvious
whether substitution was chosen because it was more economical or be-
cuase it was easier. The following paragraphs look at carbon adsorption of
trichloroethylene and have compared its cost with the cost of substituting
1,1,1-trichlorethane (methylchloroform).

Looking at adsorption (Figure 83) in terms of an equilibrium diagram,
we see that the amount of material adsorbed—in this case, trichloroethylene—is a function of both its partial pressure in the gas phase and the
system temperature.

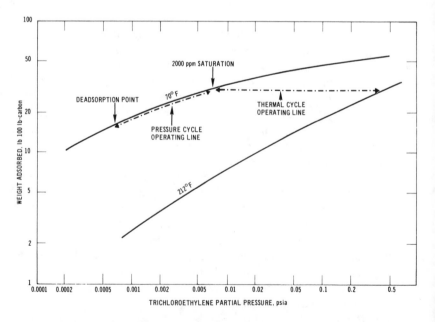

Figure 83. Estimated Isotherms for Adsorption of Trichloroethylene on Activated Carbon

For this study, it was assumed that the degreaser ventilation stream was 10,000 standard cubic feet per minute, its temperature was 70° F, and it contained 2000 ppm trichloroethylene. Two potential regeneration cycles have been considered.

The first of these is the traditional thermal cycle. In *idealized* form, carbon adsorbs trichloroethylene (TCE) until the bed is saturated. During regeneration, the bed is heated, thus increasing the partial pressure of TCE over the bed so that it can then be swept from the adsorber in a much more concentrated form. Usually, live steam is used both to heat the bed and to provide the sweep.

The second regeneration method considered is the pressure cycle. Again, carbon ideally is saturated with trichloroethylene. Regeneration is accomplished by lowering the system pressure to the point at which the partial pressure of trichloroethylene over the bed becomes large compared to the system pressure. Again, a gas stream is bled through the regenerating adsorber both to sweep the adsorber and to minimize vacuum requirements.

Pressure cycles seemed to enjoy a flurry of popularity during the early 1960s. Esso obtained some patents connected with hydrogen purification using this technique, but the pressure cycle has subsequently fallen out of style. Both were included in the study to demonstrate the relative economics of the two regeneration schemes.

Figure 84 shows a conceptual design with specifications for the pressure adsorption cycle consisting of booster fan, adsorbers, a filter to protect the vacuum pumps, two large vacuum pumps running in parallel, a condenser, a decanter, and a pump to move the recovered TCE. In the design, a choice of using superheated steam or air as the sweep gas is possible. Air would have required booster compressors with 65 pounds per square inch absolute (psia) output to increase the partial pressure of TCE to the point at which it could be condensed with water at 70° F. In addition, the condenser vent stream would have to be recycled to the adsorbers for stripping.

Because of the problems with air, the obvious sweep-gas choice was steam. The adsorber cycle allows (1) 1 hour for adsorbing, (2) 5 minutes for depressurizing, (3) 50 minutes for deadsorbing, and (4) 5 minutes for pressurizing.

Figure 85 is a conceptual flow sheet with specifications for a thermal cycle consisting of five major components: a booster blower, adsorbers (Monel* because of the chance of attack by hydrochloric acid formed in the hydrolysis of TCE), condenser, decanter, and transfer pump. As shown, the system's cycle time allows 4 hours in the adsorption mode, 2 hours heating and deadsorbing, 1-hour cooling, and 1-hour standby.

*Mention of a product or company name does not constitute endorsement by the Environmental Protection Agency.

COMPONENT SPECIFICATIONS

CODE	NAME	COMPOSITION	REMARKS
100	BLOWER	CARBON STEEL	40 hp MOTOR, 10 in. WATER-PRESSURE RISE
101, 102	ADSORBERS	CARBON STEEL	5 ft diam x 11 ft T-T, FULL VACUUM
103	FILTER		
104, 105	ROTARY VANE VACUUM PUMPS	CARBON STEEL	350 hp MOTORS, 6,000 scfm
106	CONDENSER	KARBATE TUBES CARBON STEEL SHELL	650 ft^2
107	DECANTER	EPOXY FIBER GLASS	1,000 gal
108	TRANSFER PUMP	MONEL METAL	10 gpm x 100 ft

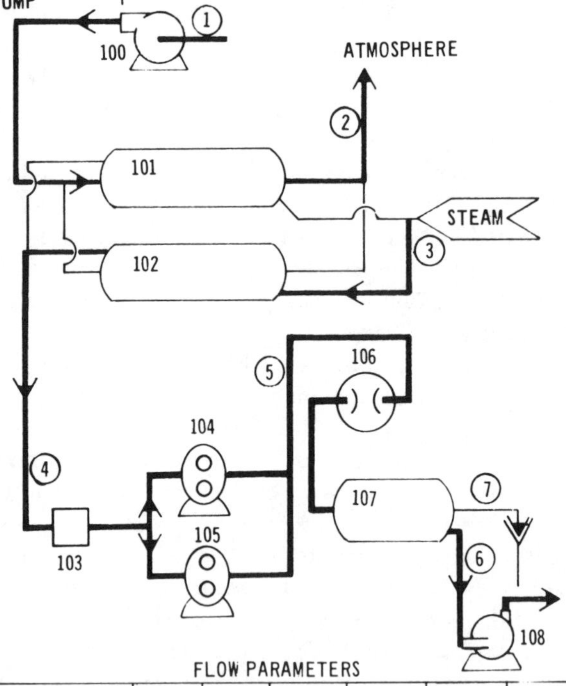

FLOW PARAMETERS

STREAM NO.	①	②	③	④	⑤	⑥	⑦
STATE	GAS	GAS	VAPOR	VAPOR	VAPOR	LIQUID	LIQUID
AIR FLOW AT 32°F, scfm	10,000	10,000	0	0	0	–	–
TRI CHLOROETHYLENE FLOW, lb/hr	93.5	9.4	0	84.1	84.1	83.5	.6
WATER FLOW, lb/hr	–	–	600	600	600	3	597
TEMPERATURE, °F	70	70	85 F	70 F	175 F	80	80
PRESSURE, psia	14.7	14.7	0.29	0.29	14.7	14.7	14.7

Figure 84. Pressure-Adsorption Swing Cycle

COMPONENT SPECIFICATIONS

CODE	NAME	COMPOSITION	REMARKS
100	BLOWER	CARBON STEEL	40 hp MOTOR, 12 in. WATER-PRESSURE RISE
101 102	ADSORBERS	MONEL METAL	6 ft diam. x 10 ft T-T, 20 psia
106	CONDENSER	KARBATE TUBES CARBON STEEL SHELL	750 ft^2
107	DECANTER	EPOXY FIBER GLASS	1,000 gal
108	TRANSFER PUMP	MONEL METAL	100 gpm x 100 ft

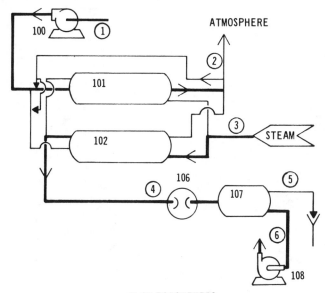

FLOW PARAMETERS

STREAM NO.	①	②	③	④	⑤	⑥
STATE	GAS	GAS	VAPOR	VAPOR	LIQUID	LIQUID
AIR FLOW AT 32°F, scfm	10,000	10,000	0	0	0	0
TRICHLOROETHYLENE FLOW, lb/hr	93.5	4.7	0	177.6	.4	88.4
WATER FLOW, lb/hr	–	–	730	730	36.2	3
TEMPERATURE, °F	70	70	212	212	80	90
PRESSURE, psia	14.7	14.7	14.7	14.7	14.7	14.7
hr/cycle hr			2/4	2/4		

Figure 85. Thermal Swing Cycle

Both cycles will meet the Los Angeles requirement of 85% reduction of
trichloroethylene emissions. The design reduction in the pressure cycle is
90%. The thermal cycle is designed to reduce emissions by 95%.

Table 21 presents an economic summary of the two processes. The pressure cycle compares rather poorly with the thermal cycle. Most of the capital- and operating-cost differences lie in the two big, expensive vacuum pumps.

Table 21. Adsorption Cycles Cost Summary

	Pressure Swing		Thermal Swing	
Trichloroethylene recovered based on 6000 hr/yr, lb/yr	500,000		530,000	
Battery Limits Plant Capital Costs, dollars (First quarter 1971)	310,000		173,000	
Operating Costs	**Use per year**	**$/year**	**Use per year**	**$/year**
Carbon, $0.40/lb	1000 lb	400	2000 lb	800
Steam, $2.00/$10^3$ lb	3600 klb	7,200	2200 klb	4,400
Cooling Water, $0.02/$10^3$ gal	1.05×10^8 gal	2,100	5.5×10^7 gal	1,100
Electricity, $0.02/Kwh	2.65×10^5 kwh	53,000	1.44×10^4 kwh	2,900
Direct Operating Labor 2 hr/shift, $4.00/hr		6,000		6,000
Supervision and indirect labor		—		—
Payroll added costs, 30% direct labor		1,800		1,800
Maintenance, 8% capital		24,800		14,000
Plant indirects, 75% direct labor		4,500		4,500
Taxes, insurance, depreciation, 1, 0.5, 9.1% capital		38,000		18,300
Interest on investment, 6% capital		18,600		10,400
Recovery cost, $/yr		151,000		64,200
Recovery cost, cents/lb recovered TCE		30.3		12.1

Assuming that a control agency requires that the degreaser discussed here must be brought under a Rule 66 type control, the economic question to be tested is: Which control scheme is cheaper, substitution or thermal cycle adsorption?

Methylchloroform losses were calculated from the following assumptions:

1. The only solvent losses occur in the ventilation system.
2. Vent losses for various solvents, on a mole basis, occur in the same ratio as their vapor pressures.

On this basis, 1,000,000 pounds of methylchloroform would be lost in comparison with 560,000 pounds of TCE from an uncontrolled system.

Oil, Paint and Drug Reporter in 1971 reported the price of trichloroethylene as 10.5 cents per pound and methyl chloroform as 14.6 cents per

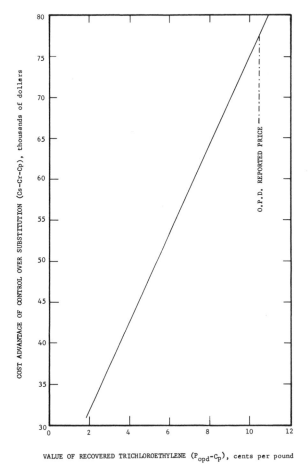

Figure 86. Effect of Additional Processing Costs on the Advantage of Recovery over Substitution.

pound. Substitution at these prices would increase the operating cost of the degreaser by about $87,000, the difference in value of the lost solvents. The cost increase that can be assigned to the TCE recovery system depends on the fate of the recovered solvent. If the recovered TCE can be reused without further processing, the cost increase is only $10,000 a year; if the recovered material must be wasted, given away (without incurring disposal costs), or reprocessed at a price equal to the O.P.D. published price, the cost increase would be about $64,000 a year.

Figure 86 shows how further processing will affect the relative cost advantage of recovery over substitution. The x-axis if the value of recovered trichloroethylene—the OPD list price minus the cost of further processing. The y-axis is the dollar savings accrued in deciding to control instead of substitute—the cost of substituting methylchloroform for TCE minus the cost of recovering TCE minus the cost of further processing.

In summary, then, this paper has shown (1) why pressure-cycle adsorption is not more commonly used and (2) why under certain conditions recovery is less expensive than solvent substitution.

Acknowledgment

I would like to acknowledge the help of MSAR project managers W. J. Cooper and A. J. Juhola who provided me with summaries of their work to help with the preparation of this paper.

REFERENCES

1. Control Techniques for Hydrocarbon and Organic Solvent Emissions, U.S. Department of Health, Education, and Welfare, Public Health Service, National Air Pollution Control Administration. Publication No. AP-68 (1970).
2. Source Inventory of Air Pollutant Emission. San Francisco Bay Area, California, Air Pollution Control District (1969).
3. Miller, F. A. "Computer Approach Increases Yields," Chem. Eng. Prog. *64* (12) 62–67 (1968).
4. Walther, J. E., and H. R. Amberg. "A Positive Air Quality Program at a New Kraft Mill," J. Air Poll. Control Assoc. *20* (1) 9–18 (1970).
5. Hydrocarbon Pollutant Systems Study. Clearinghouse for Scientific and Technical Information, 5285 Port Royal Road, Springfield, Virginia 22151.
6. Package Sorption Device System Study. Clearinghouse for Scientific and Technical Information, 5285 Port Royal Road, Springfield, Virginia 22151.

26.
CONTROL OF AIR POLLUTION FROM FERTILIZER PRODUCTION

Edward C. Bingham

Farmers Chemical Association, Inc.
Harrison, Tennessee

This is the case history of air pollution problems facing a relatively new nitrogen fertilizer manufacturing complex. The complex in question is at the Tyner Plant of Farmers Chemical Association located at Harrision, Tennessee, just outside Chattanooga. The company, which was formed in 1961, is a farmers cooperative owned by four large southeastern regional farmers purchasing and marketing cooperatives. The company built another large complex at Tunis, North Carolina, in 1969.

The Tyner plant was built in 1962 on property obtained from the federal government on a 50-year lease. The property was an unused portion of the Volunteer Ordnance Works, a government TNT plant then in a caretaker status. This property included idle industrial facilities that had been built and operated in WW II, laid away, then rehabilitated and operated during the Korean conflict, and once more laid away.

The industrial facilities provided under the lease arrangement included water pumping and treatment and distribution systems, an electrical distribution system, a steam generation plant and steam distribution system, railroads, roadways, warehouses, shops, offices, laboratories, and 10 Dupont type 55 TPD ammonia oxidation nitric acid plants with necessary compressors and auxiliaries. All of these facilities had to be rehabilitated with costs being partially offset by reduction of least payments.

Acquisition of facilities such as these obviously gave the company a running start into the then burgeoning fertilizer business, but, as will be seen later, the company really leased a number of pollution problems—both water and air.

The time period was the early 1960s; the public, including industry, had not yet been aroused to antipollution actions and clamor.

To a chemical entrepreneur and to a cost-conscious farmer cooperative group, this arrangement looked like on hard to beat. So a Girdler 250 TPD ammonia plant, a Girdler ammonium nitrate prilling and solutions plant capable of producing 350 TPD dry ammonium nitrate and 400 TPD nitrogen solutions, and a Chemico 100 TPD once-through urea solution plant were built new and integrated into the existing complex. The plant started producing in late 1962.

Since 1962, numerous changes have been made, many of which increased production capacity significantly. The ammonia plant configuration was duplicated so that it is now a 500 TPD plant. The ammonium nitrate prilling plant has been modified to produce a new high density prill. The urea plant capacity has been increased to over 135 TPD.

For an understanding of the processes involved, see the simplified flow diagram presented in Figure 87.

STEAM GENERATION

The steam generation plant acquired in the leasing arrangements was built in 1942 and had seen service in World War II and the Korean conflict. The plant contained four large coal-fired Combustion Engineering boilers, each with a capacity of 150,000 lbs. steam per hour. To serve the plant, two of these boilers were rehabilitated and were fired with good Tennessee coal in the interest of utilizing local resources. The leasing arrangement provided that steam be furnished to the government TNT plant as required. After the TNT plant was reactivated in 1965, the other two boilers were rehabilitated and fired up. Although the plant was equipped with then-modern pollution-abatement equipment when it was originally built, it emitted much soot and fly ash; pollution controls that were adequate in the 50s were no longer adequate in the 60s and 70s. The new rise in public awareness to the growing problem of environmental pollution brought in numerous complaints. Neighbors in the nearby fashionable residential community and luxury boaters urged action.

This problem was solved by the relatively simple, but costly, expedient of installing modern gas- or oil-fired steam generating equipment. At a total cost of $400,000, two 125,000 lb/hr Combustion Engineering shop-assembled boilers were installed within the battery limits of the ammonia plant to provide all steam required for fertilizer manufacture. The boilers are equipped with automatic ratio controllers and operate on either natural gas or fuel oil with completely smokeless stacks.

Auxiliary burners for gas and fuel oil were installed in the big old power house that was still operated for the government. Until a firm gas contract was negotiated, the boilers were fired with gas in the summer and fuel oil in winter.

The net results were a much cleaner environment, happier neighbors, and lower steam costs. Representative ash-fall samplings at a station located about ¼ mile downwind from the steam plants showed a significant reduction from 27 tons/sq mi/30 day before to 4 tons/sq mi/30 day after.

NITRIC ACID PLANT

Anyone familiar with the older type nitric acid plants is aware of their belching orange smoke. It's an ugly sight that is sure to bring hundreds of

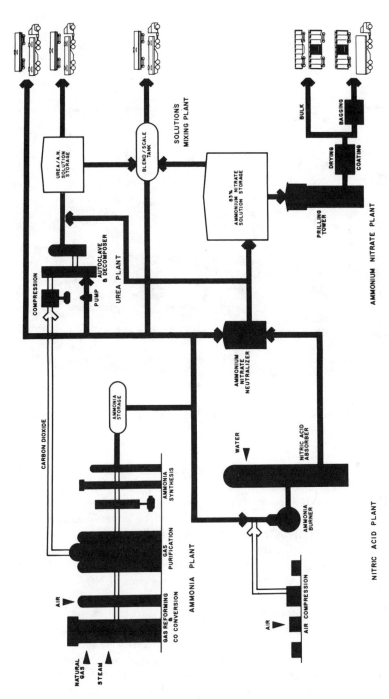

Figure 87. Flow Diagram, Farmers Chemical Tyner Plant

Figure 88. Four World War II Type 55 TPD Nitric Acid Units without Plume
        Control

pollution complaints. In addition, the NO_2 that is the orange smoke is a
health hazard at high concentrations.

The old Dupont type 55 TPD ammonia oxidation nitric acid plants that
were leased, rehabilitated, and operated emitted copious volumes of the
orange smoke (Figure 88). The smoke emission was especially heavy in the
summer when the once-through cooling water from the lake ran at high
temperatures. Because of the design of these older plants, installation of
the equipment that was becoming available to control NO_2 emissions was
impractical.

This pollution problem was dodged by a fortuitous turn of events.
About the time complaints from the public began to be registered against
the orange smoke, six of the old nitric acid plants were recaptured by the
government for use in TNT manufacture. The company entered into crash
procurement, and within 13 months, in October 1966, started operating
the second largest single train nitric acid plant in the world—a 500 TPD
Weatherly unit.

For about $50,000 additional cost, a catalytic type tail gas combustor or plume eliminator was included in the plant design. This combustor is located in the process train just ahead of the expander turbine. The combustor vessel is packed with ceramic blocks that are impregnated with a platinum catalyst. The unit was originally designed to use natural gas as fuel. However, several months' operation showed that the natural gas had a tendency to poison the catalyst. a shift to the use of the ammonia plant purge gas, which is principally hydrogen, was made and no difficulty has been encountered since.

A brief review of the chemistry of this particular pollution control measure is in order. In the synthesis of nitric acid, a mixture of ammonia and air at high temperature is passed over a platinum gauze catalyst to form nitric oxide (NO). This is followed by oxidation to the dioxide (NO_2) and subsequent absorption in water to produce aqueous nitric acid. Nitrogen and small amounts of oxygen and unabsorbed oxides of nitrogen are discharged into the atmosphere as a tail gas. This discharge represents the air pollution from a nitric acid plant. The oxides of nitrogen (NO_x) concentrations in the tail gas run about 0.1 to 0.4% by volume.

NO is colorless gas; NO_2 is an intense orange color. Decolorization of the tail gas occurs when NO_2 is catalytically reduced on the impregnated ceramic to NO in the presence of the added fuel, hydrogen, which acts as a reducing agent. Moreover, for decolorization, it is not necessary to completely reduce the free oxygen in the system.

For complete abatement, however, conversion of the NO_x to nitrogen (N_2) is required and is achieved by the addition of more fuel and the presence of the proper catalyst. The amount of fuel must be in excess of the stoichiometric requirement for complete oxygen combustion.

In this plant, the combustor is not designed to effect complete abatement; however, by using hydrogen instead of natural gas as fuel, a high degree of oxygen removal, hence abatement, is achieved.

When natural gas is used as fuel, the temperature rise ($\triangle T$) across a catalytic bed is approximately 234° F for every 1% of oxygen consumed. Conservative system design calls for an assumed light off at 850° F. Since a reactor temperature above 1400° F is undesirable, the maximum $\triangle T$ across the bed is limited to 550° F. A system operating on natural gas could, therefore, burn a maximum of 550 over 234 or 2.4% oxygen.

With hydrogen, the temperature rise is somewhat higher (a $\triangle T$ of 270° F), but the light off temperature is substantially lower (approximately 300° F). This allows for the combustion of approximately 4% oxygen. The residual oxygen in our stack runs about 1 to 3%.

The energy expended in the combustor does not all go for pollution abatement. The heat energy generated by the oxidation of the added fuel is profitably extracted by the high temperature gas expander mechanically connected to the air compressor of the ammonia oxidation system. In this plant, a steam superheater is installed between the combustor and the expander to reduce the tail gas temperature to the level required for proper operation of the turbine gas expander. The combustor also increases the life of the expander because the tail gas has no nitric acid carryover that could eventually strip out the turbine blades.

New developments toward achieving complete abatement of NO_x in nitric acid plants are being closely watched. One development involves the addition of more catalyst and a small process change. The process change involves installing a variable by-pass around the turbine gas heater where the tail gas is heated prior to entering the combustor. This change would lower the temperature of the tail gas going into the combustor below the 900° F now used. Since the tail gas must leave the combustor at 1200° F to effect reduction, a greater quantity of hydrogen fuel is required to elevate the temperature of the cooler tail gas. With a larger quantity of hydrogen (reducing agent) and a larger catalyst bed, the oxygen is exhausted from the tail gas and complete abatement is achieved. Test results on this process are being awaited.

AMMONIA PLANT

The ammonia plant, like most, is relatively air pollution free. There is some leakage and venting of ammonia, but it generally cannot be detected outside the boundaries of the complex. And, since the odor indicates loss of product, there is good economic motivation to eliminate the emission as soon as possible.

AMMONIA NITRATE PLANT

The really bad actor as an air polluter is the ammonium nitrate plant. The ammonium nitrate is formed by combining ammonia and nitric acid in a vessel called the neutralizer. Here, an 83% water solution of the ammonium nitrate is produced, and this is placed in storage as an intermediate for other products.

Alongside the nitrate plant is the once-through Chemico urea plant where an 87% water solution of urea is made by the reaction of carbon dioxide (the by-product of the ammonia plant) and ammonia. The urea solution is used in blending nitrogen solutions. Production of urea is favored by an excess of ammonia. Consisting of carbon dioxide, water, and the excess ammonia, the off-gas from the urea plant is run through the ammonium nitrate neutralizer where the ammonia is reacted with acid. The vent stack from the neutralizer then contains carbon dioxide, water, ammonia, and ammonium nitrate fines. Originally, as is the usual practice, these gases were vented into the atmosphere with no treatment.

This obviously created a severe air pollution problem. In addition, the ammonium nitrate fines were deposited on the surrounding area, dissolved

Figure 89. Ammonium Nitrate Neutralizer Vent with Wet Scrubber

during periods of rainfall, and carried into the plant wastestream, thereby aggravating the water pollution problem.

The air pollution problem was at least partially solved by the again relatively simple, but costly, expedient of installing modern equipment.

After an extensive engineering study, a scrubber was added to the vent. This scrubber uses an acidic weak solution of ammonium nitrate to scrub out the ammonium nitrate fines and the ammonia. The concentration of the scrubbing liquid is continually adjusted and the stream is recycled. A side stream is removed for product processing. About five tons of ammonium nitrate are recovered per day in this manner. This represents a considerable savings in product loss in addition to relieving part of the pollutional problems, both air and water.

The addition of the scrubber resulted in no appreciable reduction in the plume from the stack (Figure 89). The plume is used by aircraft pilots as a navigational guide in making their approach to the Chattanooga municipal airport. The neighbors, however, do not feel that it serves any useful purpose at all. The plume does not show much inclination to readily dissipate, and it has a great tendency to hug the ground and create a heavy local ground fog under certain atmospheric conditions. This is probably because of the presence of the cooled carbon dioxide gas. Also, the plume, upon mixing with other plumes (including those from the neighboring TNT plant), results in some interesting and challenging synergistic effects. This plume still poses a significant air pollution problem.

The ammonium nitrate plant was originally designed to produce what is referred to as a low density prill. Low density prills are those that are formed from a melt containing about 4.5% moisture, and the prills are made anhydrous by subsequent stepwise moisture losses in the rotary predryer, dryer, and cooler. Briefly, the low density prills were produced as follows. The intermediate 83% ammonium nitrate solution was evaporated in a calandria-type evaporator to about 96%. The vapor from this evaporator was condensed, and attempts were made to work this off in the plant's weak liquor system. But this failed and resulted in a water pollution problem.

The 96% ammonium nitrate solution was sprayed from the top of the prill tower to form the prills. From the tower, the prills passed through a rotary predryer and dryer in series. After processing through a rotary cooler, the prills were coated with a clay powder to keep them anhydrous and processed for shipment.

As the prills passed through the rotating dryers, they encountered a hot air stream. This drying air stream picked up large quantities of ammonium nitrate dust particles which exited to the atmosphere through a dust collector located on the roof. This dust collector was very inefficient and allowed fine particulate matter to collect on the roof and surrounding areas. During periods of heavy rainfall, this ammonium nitrate dust dissolved and entered the plant effluent as waste product. Also, the old process of coating with clay contributed to water pollution problems.

So it was that the low density prill process presented many air and water pollution problems as well as producing a product inferior to the high density prill that was beginning to gain favor in the industry. Possible process changes that would improve both situations were then examined.

To make a high density prill, it is necessary to evaporate the 83% ammonium nitrate solution to 99.8%, called the melt. To achieve this degree of evaporation with the calandria-type evaporator in which the vapor was condensed would produce more weak liquor than could be economically handled in the plant. The calandria-type evaporator was converted to vent instead to the atmosphere. Also, a falling film type evaporator which uses an air sweep for evaporation was added to the process. This provided the means to produce a highly concentrated melt without producing water pollution, but an air pollution problem was created with these vents (Figure 90).

Production of the new high density prills requires no predryer and dryer—only one rotary cooler, which requires relatively little dust collection. The existing dust collector adequately takes care of this and virtually

Figure 90. Ammonium Nitrate Evaporators: Falling Film Type on Left;
Calandria Type Vented to Atmosphere on Right

eliminates water and air pollution from this source. The evaporators, using atmospheric venting, create air pollution. Since making the high density prills with an additive creates self-coated prills, the use of clay coating with its attendant pollutional potential is eliminated.

In prilling low density prills, little pollution was experienced from the prill tower itself. Occasionally, some fallout of fines would be evident around the tower. This could usually be traced to dirty spray plate orifices.

Prilling of high density prills has resulted in considerable air pollution at the prilling tower regardless of the condition of the spray plates (Figure 91). Large quantities of fines are produced. A considerable quantity of these fines fall through the tower, are screened out, remelted, and recycled through the process. Another large quantity of fines is carried out of the vents in the tower by the air currents and are scattered, largely in the immediate vicinity of the tower, where they, of course, contribute to the water pollution. In addition, a heavy blue smoke is emitted from the tower during prilling. Recently, the Hamilton County Air Pollution Control Bureau has measured the equivalent opacity of the prill tower plume and found it to be a Ringelmann 3 ot 60% obscuration. The ordinance permits

Figure 91. Ammonium Nitrate Prilling Tower Showing Emissions from Top of
    Tower

only a Ringelmann 2 or 40% at present, with the code tightening to
Ringelmann 1 or 20% in 1974.

Considerable study has been brought to bear on this prill tower problem. Two recent process changes have contributed to the solution of the
problem.

When prills are formed from spraying the concentrated ammonium nitrate melt through an orifice, uniform globules are not continuously formed. Instead, the size varies considerably, and, for each regular sized prill formed, there is a small satellite prill formed; these satellite prills are thought to create the fines problem. It has been found that, when the spray plates are vibrated, prills of uniform size and without satellites will be continuously formed.

Another speculation about the prill tower problem is that the blue haze results from decomposition of the ammonium nitrate at the high temperature at which the melt is sprayed from the top of the tower. The decomposition of the ammonium nitrate is thought to be a direct function of the vapor pressure of the melt. Thus, it appears that a simple solution is merely to lower the temperature of the melt going to the top of the prill tower. Unfortunately, it's not that simple. In the first place, the melt will salt out at about 340° F or below. Also, the melt must be kept hot (about 375° F) to avoid any moisture pick up. In the past, the melt up the tower and through the spray heads has been maintained at 375° F. When this temperature is reduced to 345° F, the vapor pressure is halved. This, according to the theory, should result in a significant reduction in the blue haze. A process change was made to effect this reduction in the temperature of the melt at the spray heads. This was accomplished by jacketing the tower supply line with condensate at a controlled temperature, and some reduction of the haze has resulted.

The plant has also been faced with an air pollution problem of considerable magnitude from other emission points in the production of high density prills. The pollution sources are the ammonium nitrate neutralizer vent, the atmospheric vents on the calandria-type and falling-film-type evaporators, as well as the prilling tower (Figure 92). All of these emissions tend to produce an interesting synergistic effect. A complication was the fact that the total problem, or individual problems, had never been accurately defined by good source sampling. The nature of the emissions and the configuration of the stacks, pipes, towers, etc., from which they are discharged make for unique sampling problems. The sampling problems were evaluated by several experts who were well qualified to accomplish sampling of ordinary stacks, and they have thrown up their hands. Consequently, sampling equipment and techniques have been developed locally, and considerable data have been accumulated.

The fertilizer industry is well known for its cooperative efforts in dealing with mutual problems, especially in the area of safety. Consequently, in the fall of 1970, the company surveyed the industry on air pollution. It was found that some twenty companies producing high density prills were experiencing similar air pollution problems and source sampling difficulties.

274

Figure 92. Three Principal Emission Sources at Ammonium Nitrate Plant;
Left to Right: Prill Tower, Evaporators, Neutralizer Vent

Largely as a result of information generated at four cooperative workshops, a treatment system is now being developed in which the contaminated gas streams from the neutralizers, evaporators, and prill tower will be combined and wet scrubbed in a common scrubber. There is much evidence that the moisture from these streams will serve to agglomerate the small particles, resulting in a particle size that can be scrubbed with relative ease in a low-energy scrubber.

27.
PROCESS MODIFICATIONS FOR AIR POLLUTION CONTROL IN PULP MILLS

Sergio F. Galeano

Coordinator, Forest Products Division
Owens-Illinois, Inc., Toledo, Ohio

Many problems of air pollution control in pulp mills are suitable for solution through process modifications. The problems we refer to are provoked mainly by gaseous sulfur emissions in the chemical recovery system of the pulping process.

Process modifications should always be the first approach in any environmental control project. Not only would capital and operating costs of control-at-the-source be offset, but also the inescapable substitution of one new form of pollution (solid or liquid) for the gaseous form would be eliminated. This interrelation among the three physical forms of pollution, as commonly denominated, cannot be forgotten today, when the disposal of even dewatered sludge can result in a complicated solution.

DESCRIPTION OF THE PULPING PROCESSES

Since control through process modifications implies an intimate knowledge of the process, we will briefly describe the pulping processes involved before entering in the specifics of the examples.

Kraft Pulping

The kraft pulping process is schematically depicted in Figure 93. The main active chemicals in the kraft cooking liquor are sodium sulfide and sodium hydroxide. After cooking, the pulp is separated from the spent liquor. The spent, or black, liquor contains the remaining inorganic chemicals and organic soluble compounds from the wood.

The spent liquor is concentrated from 14 to 50% solids in multiple effect evaporators (MEE). Further concentration to 60 to 65% solids is achieved by direct contact of the previous concentrated liquor with the hot flue gases of the recovery furnace. It is in the recovery furnace that the 60 to 65% spent liquor solids is burned in a two-stage combustion process to recover the inorganic chemicals and the heat of the soluble organic compounds.

The recovered inorganics, in the form of a molten smelt of sodium sulfide and sodium carbonate, are diluted and sent to a causticizing step,

275

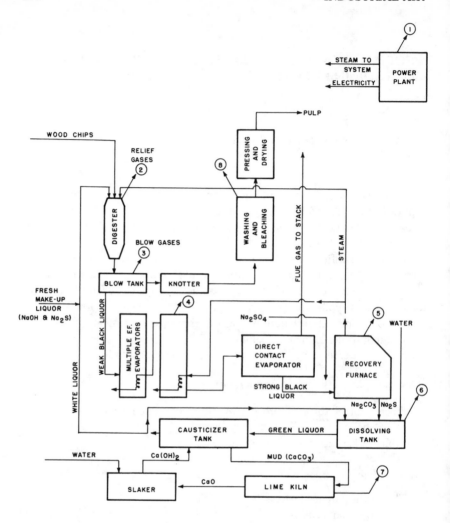

Figure 93. Typical Kraft Process

whereas, upon contact with calcium hydroxide, the cooking liquor is again formed. The causticizing and regeneration equations are:

$$NA_2CO_3 + Na_2S + Ca(OH)_2 \rightarrow 2NaOH + Na_2S + \underline{CaCO_3}$$

$$CaCO_3 \rightarrow CaO + CO_2$$

$$CaO + H_2O \rightarrow Ca(OH)_2$$

The precipitated calcium carbonate is burned in a lime kiln to regenerate calcium hydroxide after slaking.

Neutral Sulfite Semi-Chemical (NSSC) Pulping

This process, as it was practiced in Owens-Illinois mills, included chemical recovery. It was similar in many aspects to the previously explained kraft process. The main active chemical in the NSSC cooking liquor is sodium sulfite. This fact calls for a modification in the recovery system. The *sulfitation step* replaces the causticizing step of the kraft process. The reactions involved are:

$$S + O_2 \rightarrow SO_2$$
$$Na_2S + H_2O + SO_2 \rightarrow Na_2SO_3 + H_2S$$

Thus, the original sulfitation stage implied the emission of mole of odorous hydrogen sulfide for each reacting mole of Na_2S, the equivalent ot 8 to 12 lb sulfur per ton pulp according to a specific mill. Figure 94 depicts the conventional NSSC process.

With this general knowledge of the problems involved, we now turn our attention to the different process modifications.

ELIMINATION OF HYDROGEN SULFIDE EMISSIONS FROM THE SULFITATION TOWER

In the old proposed system, U.S. Patent 3,622,443, sulfur dioxide in the flue gas of a power boiler using high sulfur content fuel, was recovered by scrubbing it with an alkaline solution of sodium base. The resulting bisulfite-sulfite solution can also be formed by scrubbing the sulfur-dioxide-rich stream in the sulfitation tower. The latter step is advisable when there is no appreciable sulfur dioxide in the power boiler flue gas or when the power boilers use natural gas as fuel. The resulting solution should have a molar ratio *bisulfite/sulfite* of 4/1. This solution, once clarified of flyash if necessary, because of its origin, is then contacted with the smelt from the recovery furnace. The reactions taking place are:

Favorable:

$$Na_2S + Na\ HSO_3 \rightarrow Na_2SO_3 + NaHS \qquad \log K = 6.7$$
$$Na_2CO_3 + NaHSO_3 \rightarrow Na_2SO_3 + SO_3 \qquad \log K = 3.0$$

Unfavorable:

$$NaHS + NaHSO_3 \rightleftharpoons H_2S\ (aq) + Na_2SO_3 \qquad \log K = -0.2$$
$$NaHS + NaHCO_3 \rightleftharpoons H_2S\ (aq) + Na_2CO_3 \qquad \log K = -3.3$$

At our Big Island, Virginia, mill, where the power boilers operate on low-sulfur coal, the formation of the bisulfite-sulfite solution took place in

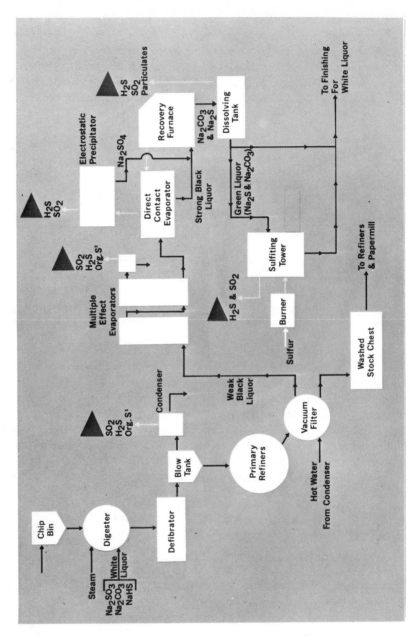

Figure 94. Conventional NSSC Recovery Process

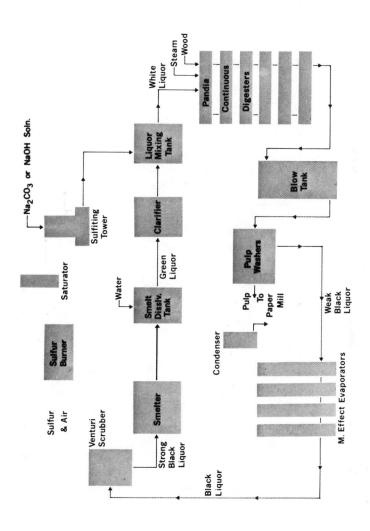

Figure 95. Modified NSSC Recovery Process at the Big Island Mill

the original sulfitation tower to make use of the sulfur dioxide source. The sulfited liquor was contacted with green liquor instead of directly with the smelt. These processes resulted in a recovery system ideally suited to that mill. Figure 95 is a schematic of main components of the system at Big Island. In this way, we were able to reduce gaseous sulfur losses in 8 to 10 lb per ton pulp.

CONTROL OF SULFUR EMISSIONS FROM THE RECOVERY FURNACE

The design of kraft furnaces has been an evolutionary process from crude burners to the new air-pollution/control-oriented units. The complexities of the different steps involved in the combustion of spent liquors demand careful investigation when considering use of pulping liquors other than kraft, especially if pollutants are to be controlled. Introduction of the new liquor in our Tomahawk, Wisconsin, kraft furnace created new technological problems, including new air pollution problems.

The operation of a pollution-controlled furnace for a kraft and/or an NSSC process must be a compromise of different and sometimes contradictory requisites. They are:

1. acceptable combustion efficiency;
2. chemical recovery as Na_2S and Na_2CO_3;
3. minimization of Na_2SO_4 formation;
4. low alkali dust load;
5. elimination of $S_4^=$ emissions;
6. low SO_2 emission;
7. control of CO in flue gas; and
8. control of nitrogen oxides.

The complex processes taking place during combustion could be simplified into a system of inorganic substances. The two existing phases in the lower zone, gas and solid, can be explained by thermodynamic postulates. Reactions and values of the equilibrium constants are available in several publications. The solution of simultaneous equations in order to obtain the partial pressures of each reacting substance at equilibrium is a rather tedious problem and is unnecessary for our approach. We estimated that a logarithmic equilibrium diagram for the system involved would clarify the problem.

Mapping the Zone of Compromise in the Equilibrium Diagram for Primary Air

Figure 96 is an equilibrium diagram for a sodium-base pulping system based on these assumptions:

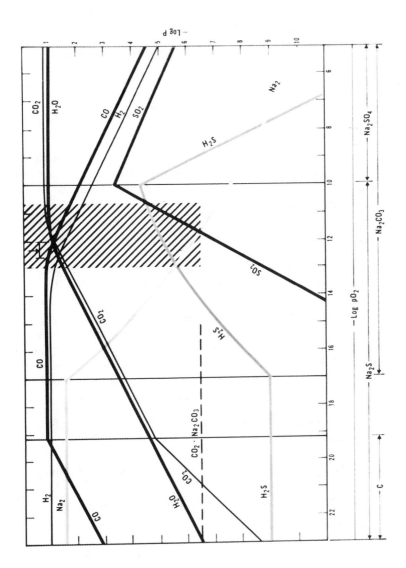

Figure 96. Logarithmic Equilibrium Diagram, Sodium-Base Liquor

$$pCO + p\,CO_2 = 0.15\,atm$$
$$pH_2 + pH_2O = 0.10\,atm$$
$$pH_2S + pSO_2 = pSO_3 + pS_2 \ll 0.01\,atm$$

Na in excess of S as to fix all S to Na and to obtain a
Na_2CO_3 residue.

Since our modification was to be achieved through air control, the
partial pressure of oxygen was taken as abscissas and the partial pressure of
the remaining compounds as ordinates, all in negative logarithms. We then
proceeded to narrow down the diagram according to our requisites until we
obtained the resulting rectangle in Figure 96. The diagram is valid for a
temperature of $1400°$ K. We selected a midpoint in which we would have
approximately 58% of C as CO_2, 42% of C as CO_2, 58% of H_2 as H_2O,
and 42% of H_2O. The amount of primary air was estimated to satisfy the
aforementioned ratios according to the elemental analysis of the spent
liquor. We also satisfied the Na_2CO_3 requirements of oxygen and kept in
mind subtracting the oxygen content of the black liquor solids.

Table 22 indicates, at different operating conditions, the final values for
primary air that were used to obtain zero ppm sulfide emissions in the flue
gases. Also shown are the values from the prediction method already
outlined. Notice how well they agree. It is possible that some additional
turbulence was required at the very low loading levels.

**Table 22. Theoretical and Real Amounts of Primary Air for Zero ppm
Sulfide Emission**

Furnace's Loading Level		Amount of Air (lb/hr)	
No. solids/hr	psig nozzle	prediction method	real*
18,600	65	61,000	63,000
17,000	55	54,000	54,000
15,600	45	48,500	49,000
11,988	35	42,000	45,500
8,800	25	36,000	43,500

*Under field conditions of zero ppm TRS (total reduced sulfur)

Secondary Air and Turbulence

The amount of secondary air was calculated by subtracting the primary
air from the total estimated requirements. Total air requirements were
ascertained in the way indicated in texts or other publications. Proper

amount and turbulence are necessary to oxidize the sulfide that will be formed in the primary zone and to guarantee proper fixation of the sulfur dioxide to the alkali present.

For our experiment, we tried to follow an approach consisting of the increase of the secondary air injection velocity. At that moment in the study, we were able to obtain zero ppm sulfide emission at the highest loading level. As the input of black liquor solids to the furnace decreased, there was a need to increase secondary air above stoichiometric amounts in order to complete the oxidation of the sulfides in our fixed black liquor solids (see Table 23).

Table 23. Stoichiometric and Effective Secondary Air

Furnace's Loading Level		Secondary Air (lb/hr)		Ratio E/S
No. solids/hr	psig nozzle	stoichiometric	effective*	F
18,600	65	53,000	53,000	1.0
17,000	55	48,000	51,000	1.05
15,600	45	38,600	48,000	1.17
11,988	35	28,000	44,500	1.55
8,800	25	16,000	43,500	2.6

*For zero ppm TRS (total reduced sulfur).

$F = 0.90 + 0.0955(P/65)^{-3} .017$

Scale of Turbulence

The correct solution for the optimum degree of turbulence needed in the secondary zone should include a scale of turbulence. The scale of turbulence will give, for each operating condition, the necessary amount of velocity of air. We like to think of it as an analogy with the von Karman-Howarth equation for the second order correlation function $f(r,t)$. If r is the distance between two points, a correlation between the fluctuations at these two points will give a measure of turbulence. Another analogous element can be introduced if we consider the secondary ports system as a grid or mesh producing the turbulence. The decay of the intensity of turbulence is also expressed by the function $f(r,t)$.

For our fixed geometric dimensions in the furnace, a decrease in the loading will bring about a decrease in the distance, r, and time, t, after the grid, and, consequently, a decay in turbulence. For lower loadings to the furnace, it will be necessary to increase turbulence in the same fashion it decays. In other words, we need to trace back the curve which gives $f(r,t)$.

Figure 97 indicates the high loading point (1,1) and the reverse curve (dotted line). Point (.4,2.5) was determined by proportion and some

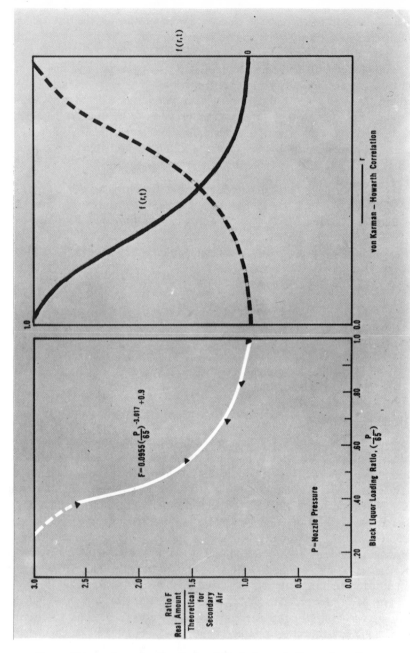

Figure 97. Analogy for the Scale of Turbulence in Secondary Zone

trial-and-error. With these two points, an operation equation was developed and estimations for each operation level were made.

REDUCTION OF SULFUR DIOXIDE EMISSIONS FROM THE RECOVERY FURNACE (KRAFT AND/OR NSSC)

Sulfur dioxide emissions from the recovery furnace in both processes can be generally high, above 500 to 700 ppm. This situation can be corrected without any source control equipment just by an understanding of the process. To recombine the sulfur dioxide with the alkali, in the upper or oxidation zone there is, first, need for an adequate amount of alkali. Thus, a relationship bewteen the Na/S ratio by weight in the spent liquor and the sulfur dioxide emissions can be easily expected. Furthermore, when total reduced sulfur (TRS) emissions are decreased, there is a tendency to further decrease the original Na/S ratio. Since the system will tend to balance itself, the original Na/S ratio will be reestablished at the expense of additional SO_2 losses. Figure 98 illustrates the results of experimental data, collected during our studies, which clearly indicates this relationship.

CONTROL OF ODOROUS EMISSION IN THE KRAFT PULPING OF SOUTHERN PINE

The spent liquor of the kraft process has a sodium sulfide concentration ranging between 7 and 12 gpl at 13 to 14% solids. The liquor is concentrated in multiple effect evaporators to 45 to 52% solids. The liquor is further concentrated to 60 to 65% solids by contacting it with the hot flue gases of the recovery furnace in direct contact evaporation. Odorous emissions result from the stripping of organic and inorganic sulfur compounds at the direct contact evaporator. This is the conventional kraft recovery system. The stripping reaction is given by the equation:

$$Na_2 + CO_2 + H_2O \rightarrow Na_2CO_3 + H_2S$$

A solution to this problem is liquor stabilization by oxidation with air according to

$$2Na_2S + 2O_2 + H_2O \rightarrow Na_2S_2O_3 + NaOH$$

This stabilization should be accomplished as soon as the spent liquor and the pulp are separated. The foaming characteristics of southern pine spent liquors make it very difficult to oxidize the weak liquor with air. Only after it has been desoaped and concentrated in the MEE, has it been possible to oxidize southern pine kraft liquor with air.

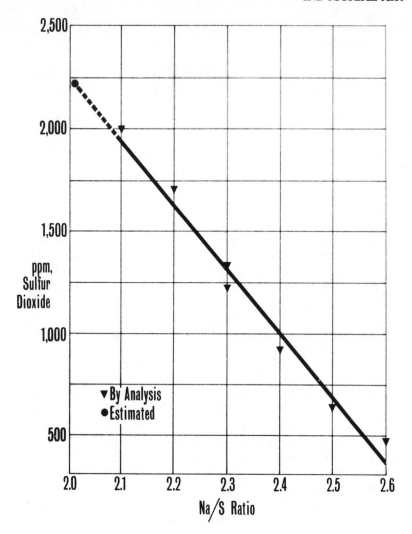

Figure 98. SO_2 Reduction with Increased Na/S Ratio

We have been studying for some time a black liquor oxidation system able to accomplish the following objectives: (1) no foam formation, (2) minimum capital cost and equipment complexity, (3) great flexibility for installation in any existent pulp mill, and (4) improvement over the known oxidation reactions taken place in conventional air oxidation systems, particularly reversion.

One of the drawbacks of the conventional black liquor oxidation systems with air is the reversion phenomenon. The word "reversion" has been used and misused for many years to label the lack of fulfillment of

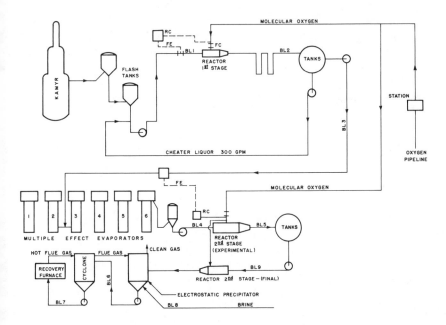

Figure 99. General Schematic of the MOBLO System

the expectations placed on the oxidation of the black liquor. In its proper context, it should be applied only to the increase of sulfide content in the liquor once it has been properly oxidized. Proper oxidation can be defined as that conducted at a proper temperature to avoid elemental sulfur formation and given enough time for completion of the reaction to thiosulfate.

Description of the Moblo System at Orange

The use of molecular oxygen for black liquor oxidation, the MOBLO system, can be fashioned in different operational variants according to a specific kraft process. Figure 99 illustrates the variant used at Orange. Since this mill has Kamyr digesters, the first stage tubular reactor follows the flash tank pumps. Were it a kraft mill with batch digesters, the first stage could be located after the weak liquor storage tanks, with provisions for adequate liquor temperature.

The tubular reactor consists essentially of a reactor and the continuing pipeline. The oxidized liquor is stored in different tanks where some additional oxidation still takes place. The oxidized liquor is concentrated from 14 to 16% solids to 50 to 53% in the multiple-effect evaporators. An experimental second-stage reactor similar, but not equal, to the first stage one was installed after the multiple-effect evaporators.

The control system for the first stage has been depicted in U.S. Patent No. 3,655,343. The control for the second stage is very similar to it, the exception being that the liquor flow signal is taken from the existing flowmeter at the feeding of the evaporators.

Working Equations in the First Reactor

It is possible now to predict with a higher degree of accuracy the operation of the first stage tubular reactor. Two working equations are used. The inputs for the equations are:

Q_l = liquor flow rate in gallons per minute
Q_g = oxygen flow rate in standard cubic feet per hour
C_l = initial concentration of sodium sulfide in gpl
C_f = final concentration of sodium sulfide in gpl
L = length of pipeline in ft.

The first working equation is the one giving the value of the rate constant,

$$k_1 = 0.1038 + 0.1421 \, (Q_g Q_1) 10^{-6}$$

From it we substitute in,

$$C_f = C_1 e^{-k_1} \, (150/.243 Q_1)$$

for the condition of 150 feet of 10-in. diameter pipe. This is essentially the same type of equation developed from the original mill trial.

In order to ascertain the reasons for the observed differences in value of the reaction rate constant, a parameter of turbulence, the product of the fluid flow rates, was developed. Figure 100 illustrates the plot of the different paired values of turbulence parameter versus reaction rate constants. Points illustrated as squares represent corresponding values from the original trial. A regression equation permits estimation of the rate constant at any operational condition.

Working Equations for the Second Reactor

In the first-stage reactor, the assumption of a first-order reaction in sulfide was made. The formula of the type $C_f = C_1 e^{-kt}$ was again used. The reaction proceeds slower than in the first stage, but there was need to find out the reasons for the wide range of rate constant values.

As in the first stage, we studied the dependence of the rate constant with the degree of turbulence. The same type of turbulence parameter was chosen: the fluid flow rates product. Figure 101 illustrates the plotting of k_2 *versus* the turbulence parameter.

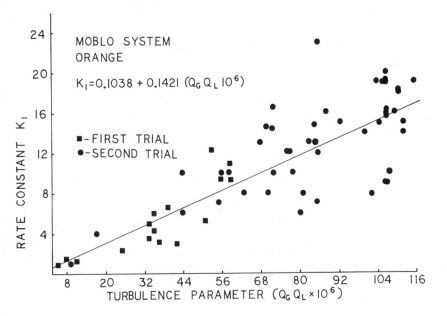

Figure 100. Rate Constant *versus* Turbulence Parameter First Stage

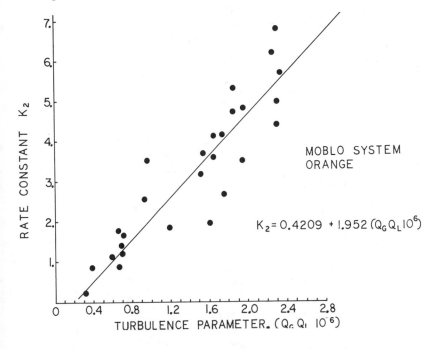

Figure 101. Rate Constant *versus* Turbulence Parameter Second Stage

Provisional working equations can be advanced for the second-stage reactor, bearing in mind that it needs to be recalculated for any new reactor. Components are as in the first-stage reactor equation, and the expressions are as follows:

$$k_2 = 0.4209 + 1.952 \quad (Q_g Q_l)10^{-6}$$
$$C_f = C_l e^{-k_2} \quad L/(.667Q_1)$$

L is the total length of the tubular reactor; pipe diameter is 6 in.

Reversion and Its Kinetic Predicting Formulation

The MOBLO system has had a peculiar characteristic; under the conditions at the Orange Mill where the system has been installed since 1971, no reversion has been detected in the oxidation of the weak liquor. Nevertheless, upon concentration in the evaporators, reversion has been noticed, and it was also noticed after the touch-off second reactor.

A Proposed Kinetic Formulation to Predict Reversions,

Studying the data on reversion for the two conditions mentioned above and with the observation that concentration (percent solids in the black liquor) is a relevant factor, we developed different equations linking initial and final sulfide concentration, percent solids, and time after oxidation or concentration. A suitable expression is given by:

$$C_t = - \left[0.06 \ (\% \ \text{solids}) - C_1\right] e^{-0.06t} + \left[0.06 \ (\% \ \text{solids})\right] - 0.06t_1 e^{0.01t}$$

$$t_1 \rightarrow 8.12 \ C_o$$

C_t is the sodium sulfide concentration at any given time after oxidation, t is the time in hours, C_1 is the initial sulfide concentration after oxidation, C_o is a pseudo oxygen concentration in the liquor (in gpl), as described in Table 24. Finally, t_1, is the estimated time for reversion to level off.

Table 24. MOBLO System—Orange Kinetics of the Reversion Calculation of C_o

Steps:

1. Find $O_2/Na_2 S$ molar ratio.
2. Find initial $Na_2 S$ concentration before reactor gpl.
3. Estimate flow to reactor in g mole, No. 2/78.
4. Initial O_2 concentration No. 3 x No. 1 in g molds/liter.
5. Initial O_2 concentration in gpl, No. 4 x 32, before any reastion.
6. % of O_2 used from data taken on sulfide converted.
7. Calculated concentration of O_2, No. 6 x No. 5, gpl.

Table 25. MOBLO System—Orange Kinetics of the Reversion Phenomenon
Predicted vs Observed Values

Code and Date	Time hrs.	$Na_2 S$ gpl	O_2 gpl	Predicted $Na_2 S$	Observation
BL5S	0	0.006	2,45	0.006	oxidized S.B.L.
2/19/72	3.25	0.096		0.349	
	8.42	0.255		0.720	
	24.25	1.099		1.14	
RTSB	0	0.163	2.837	0.163	oxidized S.B.L.
2/22/72	2.25	0.390		0.386	
	4.25	0.43		0.55	
	5.5	0.52		0.64	
	7.75	0.65		0.79	
	12.17	0.76		0.98	
	20.25	1.14		1.15	
	22.47	1.32		1.28	
RTSBL	0	0.006	2.30	0.006	oxidized S.B.L.
2/22/72	1.0	0.071		0.12	
	5.0	0.460		0.49	
	15.6	0.978		1.01	
	19.5	1.02		1.13	
RT2SL	0	1.723	0.54	1.723	
	1.42	1.702		1.741	
	4.92	2.16		1.80	
	6.42	2.13		1.88	
	22.42	2.64		2.45	
	27.92	2.78		2.56	
RT2ST	0	1.463	0.54	1.463	oxidized S.B.L.
	1.83	1.56		1.514	
	3.08	1.73		1.541	
	23.5	2.75		2.41	
RT5SL	0	.606	1.12	0.606	oxidized S.B.L.
	2.1	.869		0.763	
	4.0	.972		0.88	
	6.2	1.273		0.99	
RTWBL	0	0.215	0.657	0.215	oxidized W.B.L.
	1.0	0.156		0.177	
	5.0	0.068		0.025	
	19.5	0.156		0.211	
RT2WL	0	0.1022	1.05		oxidized W.B.L.
	1.25	0.039		0.063	
	4.75	0.078		0.031	
	25.75	0.078		0	
RT3WL	0	1.113	1.35	1.113	oxidized W.B.L.
	.25	1.098		1.098	

Figure 101 indicates how well the prediction equation expresses the reversion as observed from the experimental data gathered. The model also fits the conditions of no reversion and continuing oxidation observed in the oxidized weak liquor. Table 25 shows only a few examples, but many others were analyzed with similar results. Data supplied by Champion International from the Pasadena Mill reveals reversion on a totally oxidized strong liquor with air. This liquor was oxidized only while strong (35 to 40% solids).

The role of solids concentration as the promoter of reversion of the already oxidized weak liquor upon concentration (or of the would-be oxidized strong liquor) is very much stressed by this information as a key factor in reversion.

CONCLUSIONS

After these examples, it is evident that much can be accomplished in reduction on elimination of air pollution by process modification. Our own experience also shows that almost all systems possess an in-built capacity to assimilate changes without detracting product quality. As an example of the above, we have recently announced a nonsulfur pulping process for our semi-chemical mills, in operation since the beginning of 1972.

REFERENCES

Galeano, S. F. U.S. Patent 3,655,343, April 11, 1972.

Galeano, S. F. U.S. Patent 3,622,443, November 23, 1971.

Galeano, S. F., and C. D. Amsden. "Oxidation of Kraft Weak Black Liquor with Molecular Oxygen," *TAPPI* 53 (11), 2142(November 1970).

Galeano, S. F., D. C. Kahn, and R. A. Mack. "Air Pollution Controlled Operation of a NSSC Recovery Furnace," *TAPPI* 54 (5), 741 (May 1971).

28.
CONTROL OF ATMOSPHERIC EMISSIONS IN THE KRAFT INDUSTRY

Russell O. Blosser

National Council of the Paper Industry
  for Air and Stream Improvement, Inc.
New York, N.Y.

More than 80% of the chemical pulp produced in the United States is manufactured by the sulfate or kraft process. The dominance of this process, its characteristic reduced sulfur emissions resulting from the use of sodium sulfide as a pulping chemical, and its subsequent recovery account for the emphasis placed on air quality protection from this pulp manufacturing process.

The kraft pulping process consists of several unit processes, most of which are responsible for reduced sulfur emissions that differ both in magnitude and composition. These unit processes consist of digestion, pulp washing, cooking liquor evaporation, recovery furnace system, smelt tank exhaust, lime kiln, lime slaker vent, and the exhaust gases from black liquor oxidation.

Significant changes have occurred in the available knowledge concerning kraft reduced sulfur emission control capability and its application. Most of these changes have come about in the past five years. Probably the most significant of the events relative to control technology development is the definition of several furnace operating conditions that relate to control of emissions from the kraft recovery furnace system. Simultaneously, the potential emission control capability of black liquor oxidation as it relates to flue-gas direct-contact evaporator emissions has been more extensively explored, appreciated, and partially demonstrated. Prior to this time, a satisfactory definition of the emission characteristics and a satisfactory control technology were developed for the digester and the multiple-effect evaporator system.

Coincidental with the latter stages of investigation of the kraft recovery furnace system, an extensive field study program was begun for the emission sources that were heretofore considered as miscellaneous. This program had as its objectives a definition of the reduced sulfur emissions and the identification of process and operating conditions which could be related to these emissions from the smelt dissolving tank, brown stock washer system, black liquor oxidation tower, lime mud washer, and lime slaker vents.

Finally, field studies aimed at defining the factors responsible for odorous reduced sulfur emissions from lime kilns were also begun.

DIGESTORS AND MULTIPLE-EFFECT EVAPORATOR CONTROL TECHNOLOGY CAPABILITY

The reduced sulfur emissions from digestion of wood chips and the subsequent evaporation of the black liquor prior to recovery of the cooking chemicals are generally classified as low-volume, high-concentration sources. They characteristically contain significant amounts of the higher molecular weight reduced sulfurs, dimethyl sulfide and dimethyl disulfide, neither of which is practically treated by chemical means. Burning or thermal oxidation either after mixing with the combustion air for the lime kiln or in a separate incinerator is the most extensively used control technique although some of the off gases are treated with chlorine. The systems used for thermal destruction of these emissions are described elsewhere.[1] Regardless of the method used for burning, properly designed systems result in essentially complete oxidation of the reduced sulfur compounds from these sources to sulphur dioxide.

KRAFT RECOVERY FURNACE SYSTEM CONTROL TECHNOLOGY CAPABILITY

The recovery furnace system represents potentially the largest and most variable of emission sources in the total kraft recovery system. One means of emission control has involved the elimination of a unit process, the flue gas direct contact evaporator, which has been a portion of the recovery process almost since its inception. Of perhaps equal, if not greater, importance has been the progress made in identifying those factors responsible for achieving minimal odorous reduced sulfur emissions within the furnace itself.

An extensive series of studies of 26 kraft furnaces, each operated at varying black liquor solids firing rates, dry-solids:steam-production ratio, combustion-air:dry-solids ratio, and excess oxygen levels, has been reported elsewhere.[2] To summarize briefly, the studies showed the following: High residual oxygen is not in itself a universal indication of minimum emission rates since such minimum levels were observed at excess oxygen values of from 2 to 4%. Similarly, some furnaces operated in that range yielded reduced sulfur emissions of over 50 ppm, reflecting the importance of such factors as furnace gas turbulence and residence time and air distribution to the various firing zones. The combustion-air:dry-liquor-solids ratio generally fell between 3.5 and 4.5 for minimal emission, being influenced by methods of air distribution and BTU content of the liquor solids. Smelt bed height increase above secondary air ports was found to be reflected in increased emissions, while smelt falling from the furnace walls produced sharp temporary increases in emission levels.

A general correlation between both volumetric and cross-sectional firing rates and minimum emission levels was noted. To achieve reduced sulfur levels of 5 ppm or less generally required loadings of less than 1.8 lbs per ft^3/hr and 130 lbs per ft^2/hr, respectively. A significant number of furnaces, however, achieved this level of emission at substantially higher loadings. This underscored the importance of individually establishing the emission control capabilities of each furnace through such test programs, and the lack of justification for regulatory agency imposition of furnace loading limitations. In this regard, in only one instance did the minimal reduced sulfur emission level coincide with the nominal, or nameplate, rating of a furnace. This level was encountered commonly at 115% of nominal rating, and in some instances at as much as 140% of that rating.

A major phase of this study, and one not as yet completed, dealt with establishing minimum emission level changes as a function of black liquor oxidation during the passage of furnace flue gas through the contact evaporator. As a result, the emission control capability resulting from oxidation of sodium sulfide in black liquor with air or oxygen to form a more stable form of sulfur, hence reduce or eliminate sulfur stripping upon contact with acidic combustion gas, is not fully recognized. It appears now that completion of black liquor oxidation to well below 1 g residual Na_2S per liter at the point of liquor introduction to the direct contact evaporator is required to lower the reduced sulfur gas content uptake to less than 2 ppm. Only a limited number of existing oxidation systems have been shown to operate to this level, indicating the need for additional studies and refinement of oxidation systems through addition of polishing features and improved process control. The effectiveness of adequate black liquor oxidation is illustrated in Figure 102 where changes in the flue gas total reduced sulfur (TRS) composition as a function of residual sodium sulfide in black liquor is shown.

INTERPRETATION OF RECOVERY FURNACE SYSTEM STUDY FINDINGS

It appears at this time that overall recovery furnace system reduced sulfur emissions can be brought to the 5 ppm level or 0.2 lb/ton level regardless of the choice of system design approach, namely, those with and without the direct flue gas contact evaporator. When viewed against the not uncommon emission level for the overloaded and completely uncontrolled furnace system (*i.e.*, absence of black liquor oxidation) of up to 1000 ppm, the performance improvement potential of at least 99.5% puts the remaining level as well as reported differences between the two system approaches in a more realistic perspective.

Coupled with (1) the fact that introduction of highly oxidized black liquor into the direct contact evaporator permits it to serve as an alkaline scrubber, absorbing any intermittent surges in furnace gas sulfide content,

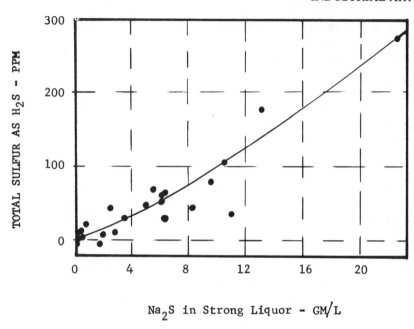

Figure 102. Total Sulfur Increase across the Direct Contact Evaporator

and (2) the existence of as yet incompletely resolved operational problems for the newer furnace system designs, the factors of choice between the two basic approaches remain more complex than was originally believed.

Ideally, downwind effects on human odor receptors should form the appropriate basis for selection of alternate furnace systems differing only minimally in reduced sulfur emission potential rather than administratively selected emission standards. The duration of exposure and permissible exposure concentrations are major elements in such problem analysis. The reliable measurement of human response to an odor is a classical example of the difficulties encountered in precise application of the "receptor response-corrective action need" methodology to problem solving.

The estimation of frequency and duration of ground level concentrations of an emission component, such as the reduced sulfur compounds, is possible using existing meterological predictive techniques. These represent the soundest approach available for linking emission rate to expected ground level concentrations. To complete the spectrum of informational needs for defining kraft recovery system control technology requirements is an understanding of the nature and magnitude of emissions from several unit processes, some with multiple emission points. Among these are a series of miscellaneous sources which include vents on brown stock washing systems, black liquor oxidation systems, smelt tanks, lime mud washer and slaker vents, tall oil plants, and lime kilns as well as others.

MAGNITUDE OF MISCELLANEOUS REDUCED SULFUR EMISSION SOURCES

A series of field investigations has defined the range of emissions from several of these sources and has related process operating variables to the emission rate.[3] The TRS (total reduced sulfur) emission from 23 total washer system vents ranged from 0.01 to 0.84 with a median of 0.16 and a mean of 0.28 lbs/ton pulp. Approximately a third of the total originates from the roof vent system and the remainder from the undervent system (*e.g.*, seal tank, foam tank, or vacuum pump exhaust). The sulfur emissions were exclusively organic sulfides and are dominantly the heavy organic sulfur compounds dimethyl sulfide and dimethyl disulfide.

The most significant operating variable associated with the emissions from this source is the use of contaminated condensates, *e.g.*, evaporator condensates, for wash water. The emissions from the total system are almost 150% greater when condensates are used as wash water, 0.19 compared with 0.46 lbs/ton pulp. The most significant increase was noted at the roof vent system where losses are 4 times as great, or 0.06 compared to 0.24 lbs/ton pulp. The bulk of the increase was at the hoods venting the washer receiving fresh condensate.

Pulp consistency in the blow tank (an indirect measure of agitation and stripping at this point, hence loss to the low-volume noncondensable system) accounts for some differences in emission rate when fresh water is used at the washers. The volume of vent gas per unit or production, an indication of the degree of agitation in the seal tank, was found to be closely related to the emission rates from the undervent system; the greater the gas flow, the greater the emission. These data are shown in Tables 26 and 27.

Table 26. Summary of Kraft Mill Brown Stock Washer Reduced Sulfur Emissions

TRS, lb ton Pulp

	min.	max.	median	mean
30 Roof vent systems	0.01	0.60	0.05	0.10
23 Undervent systems	0.01	0.63	0.11	0.16
23 Total washer systems	0.01	0.84	0.16	0.28

Table 27. Effects of Pulp Washer Water Source on Reduced Sulfur Emissions

TRS, lb ton Pulp

Wash Water Source	roof vents	under vents	total system
Contaminated condensate	0.24	0.22	0.46
Fresh water	0.06	0.13	0.19

In Table 28 are summarized the black liquor oxidation system vent emissions from eight week liquor and seven heavy liquor oxidation systems. These, like the sulfur emissions from the brown stock washer system, are essentially all high molecular weight organic sulfur compounds. The sulfur emissions from the weak liquor oxidation systems ranged from 0.02 to 0.22 with a mean of 0.11 lbs/ton pulp. The bulk of the data are grouped close to the mean with the high accounted for in a system where brown stock washer undervent systems were low. The extremes and mean sulfur emission rates from heavy liquor oxidation systems are almost identical with those from weak liquor oxidation systems. The lowest emission rates were measured from the second in a series of oxidation units, or the polishing unit, while the highest were from high-performance single stage units being fed liquor with the highest initial sulfide concentration. These data are shown in Table 28.

Table 28. Black Liquor Oxidation Vent System Emissions

TRS, lb ton Pulp

Type of System	min.	max.	mean
Weak liquor	0.02	0.22	0.11
Heavy liquor	0.01	0.18	0.10

The sulfur emissions from smelt dissolving tank vents are shown in Table 29. No significant difference was noted between those with and those without particulate emission control devices, the median being 0.01 and 0.015 lb/ton pulp respectively. The emissions at this source were found to be almost exclusively hydrogen sulfide when either fresh water, weak wash from the lime mud clarifier, or lime mud washing filtrate was used as the liquid medium in the particulate emission control device. In two situations, the scrubbing solution used on the particulate control device was found to contribute to the emission level. Evaporator condensates, when used as scrubbing solutions, contributed organic sulfides. In some systems where lime mud clarifier supernatant was used as a scrubbing solution, a small amount of hydrogen sulfide was generated in the scrubber. This was attributed to mild acidification of the scrubbing solution.

Several other miscellaneous source emissions have been measured. Some of these have arbitrarily been described, due to their nature, as low-volume, high-concentration sources. They include vents associated with continuous digesters (chip filler and pocket relief valves), diffusion washers, blow heat condensate handling systems, and turpentine decanters. The sulfur compound concentrations in these emissions ranged from 100 to 4000 ppm with emission rates, where flow could be measured, from 0.05 to 0.4 lb/ton.

Table 29. Summary of Smelt Dissolving Tank Vent Emissions

TRS, lb ton Pulp

	range*	median	mean
18 Vents with particulate control	nd = 0.12	0.01	0.025
6 Vents with no particulate control	nd = 0.11	0.015	0.025

*nd = not detectable.

Other process vent emissions which were measured and judged not to be of importance, due to either (1) the low concentration of reduced sulfur compounds present, less than 1 ppm, or (2) their low volume, included vents on lime mud washer hoods, lime slaking systems, board machine couch pits, and refiner reject tanks.

The lime kiln study has not advanced far enough to warrant a detailed discussion at this time. Operating variables which have been correlated with sulfur emission levels include residual sulfide content of the lime mud, cold end kiln temperatures, and residual sulfur content of the scrubber water.[4]

MISCELLANEOUS SOURCE EMISSION CONTROL

The sulfur compounds in miscellaneous emission sources resulting from the agitation and stripping of black liquor characteristically and almost exclusively are high molecular weight compounds. The practical possibility of their capture by reaction with other chemicals is limited. Caustic solutions which find traditional application in reactions with low molecular weight acidic sulfur compounds are not effective in the capture of the high molecular weight sulfur compounds present. Where these miscellaneous source emissions are treated before discharge, they are burned. In some cases, the volume of gas is sufficiently low to incorporate it with the low-volume gases from digestion and evaporation for destruction in the kiln or auxiliary incinerators.

The only existing process systems possibly capable of handling the high-volume sources are the recovery furnace and power or steam generation boilers. Engineering practicalities, excessive cost to convert existing furnaces or boilers, and the potential danger of lowering furnace smelt bed temperature sufficiently to cause excessive SO_2 generation in the recovery furnace must all be considered in selecting this method of disposal. The gas volumes are, however, so great as to make separate incineration of high-volume sources without heat recovery an impractical exercise.

SIGNIFICANCE OF MISCELLANEOUS SOURCE EMISSIONS IN PLANNING AIR QUALITY PROTECTION PROGRAMS

Some observations can be made concerning the possible significance of these miscellaneous source emissions in planning air quality protection programs. First is the interrelationship of ground level concentrations, the magnitude of these sources, and their normal method of discharge compared to the kraft recovery furnace system emissions. The second is concerned with their composition, which places practical limits on technology applicable to their effective control.

As an example, assume that reduced sulfur emission control technology capability for new kraft recovery furnace systems averages around 0.1 lb/ton pulp (about 3 ppm). Assume further that all low-volume, high-concentration non-condensible streams are treated by some means equal in effectiveness to burning, the other miscellaneous source reduced sulfur emissions can equal or substantially surpass those of the recovery furnace system. Within the structure of existing knowledge, they are not likely to be less when control of known process operating variables is practiced.

The vent gases from these systems are normally reasonably cool and are discharged at a reasonably low velocity at a relatively low elevation, all factors which inhibit their dispersion and mixing in the ambient air. Their contribution to the ground level concentration of reduced sulfur as the result of a specific mill discharge is a function not only of the amount and their method of discharge but also of the topography and meterology of the area.

To understand the interrelationship of all these factors for specific situations requires the use of meteorological predictive techniques. There is sufficient information available from special studies already completed to illustrate that both the magnitude and the method of discharge of miscellaneous source emissions deserve attention in estimates made to predict ground level concentrations and possible public response to their presence.

Of equal and possibly greater significance is the consideration of miscellaneous source emissions in control program development at existing mills. For example, consider a situation where the only recourse for reducing emissions from a kraft recovery furnace system whose emission control capabilities are no less than 0.5 lb/ton pulp is a new furnace. Equally as much improvement in ambient air quality may be accomplished by reduction in miscellaneous sources as that achieved with improved recovery furnace emissions. In this situation, proximity of the mill to the public, methods of discharge of the gases, and local meteorology may suggest improved methods of miscellaneous source emission disposal as a logical alternative course.

SUMMARY

The extensive work of the past five years has produced substantial improvements in kraft pulping process sulfur odor emission control capability. This is most simply and dramatically illustrated by a comparison of previously uncontrolled emission levels with those generally achievable with present control capability (see Table 30). For uncontrolled kraft recovery furnaces, the mean was 9 lb/ton and ranged as high as 30 lb/ton. Similarly, the uncontrolled contact evaporator contribution to emissions can amount to 10 lb/ton. With well-devised control, these can be reduced to 0.1 and 0.1 to 0.15 lb/ton, respectively, with a number of systems of various ages currently operating at 0.5 lb/ton. Other uncontrolled source emissions which can amount to 6 to 10 lb/ton can be controlled at 0.25 lb/ton or less. There is every reason to be optimistic that the odorous conditions long associated with this industry by the adjacent public can be effectively dealt with, short of complete odor emission elimination at all process sources. It should be understood, however, that remedial measures at existing mills will of necessity be complex and costly; therefore, the consideration of alternative control strategies now shown to be available is justified.

Table 30. Control Technology Capability in the Kraft Process

	Lbs Reduced S/ton	
	no control	control
Recovery Furnace	9	0.1
Contact Evaporator	10	0.1–0.15
Lime Kiln	1–2	0.1–0.2
Digesters and Evaporators	2–7	0
Brown Stock Washing Systems	0.8	0
BLO Systems	0.1	0
Smelt Tanks	0.1	0.01

It is, of course, no mere coincidence that the progress described above has been paralleled by increasing regulatory attention and emission standard setting activities. It is to be hoped, therefore, that, precisely at the point where major improvements are possible through selection of alternative strategies coupled with use of meteorological techniques for predicting dispersion and resultant downwind conditions, administrative regulations will embody the necessary flexibility to permit full use of such knowledge in designing individual mill control programs to achieve the desired goals of odor-nuisance-free ambient air in adjacent communities or areas having significant human presence.

REFERENCES

1. NCASI Air Quality Improvement Technical Bulletin No. 34, "Current Practices in Thermal Oxidation of Non-Condensible Gases in the Kraft Industry" (October 1967).
2. NCASI Air Quality Improvement Technical Bulletin No. 44, "Factors Affecting Reduced Sulfur Emissions from the Kraft Recovery Furnace and Direct Contact Evaporator" (December 1969).
3. NCASI Air Quality Improvement Technical Bulletin No. 60, "Factors Affecting Emission of Odorous Reduced Sulfur Compounds from Miscellaneous Kraft Process Sources" (March 1972).
4. NCASI Special Report No. 71-01, "Suggested Procedures for the Conduct of Lime Kiln Studies to Define Minimum Emissions of Reduced Sulfur through Control of Kiln Scrubber Operating Variables" (January 1971).

29.
CONTROL OF ATMOSPHERIC EMISSIONS FROM THE MANUFACTURING OF FURNITURE

Charles F. Sexton, Jr.

Buffalo Forge Company
Knoxville, Tennessee

In the furniture industry dust is generated by such machines as saws, moulders, planers, shapers, augers, floor sweepers and sanders. Although not producing the largest volume of dust, sanders, in most cases, produce the largest number of particles, the smallest size particles, and, therefore, the hardest particles to collect.

In practically all woodworking plants today, the machines are equipped with hoods which capture the dust. Good hood design is very important because the lower the air volume, the lower the overall cost of the dust collection system. Thus, a good hood design strives for maximum efficiency at minimum air volume. The volume of air handled is directly proportional to the area of the hood opening and the capture velocity. Capture velocities may vary from 100 feet per minute to more than 2000 feet per minute. Although most dust sources are already hooded, economics and efficiency dictate that these present hoods be considered for redesign prior to installation of high efficiency dust collection equipment.

The dust is conveyed from the hoods in ducts (or pipes) which are sized for equal suction at all hoods. The most common methods for sizing ducts are constant velocity, static regain, equal friction, and velocity reduction. In some existing duct systems, blast gates (or dampers) are being used for balancing, although these are not recommended because plugging of the ducts very often accompanies their use.

Velocities in ductwork should be maintained at 2500 to 4500 feet per minute or higher to insure continuous conveying of the material, and the air flow should be 50 cubic feet per pound of material or more, with cleanout doors located in the ductwork, particularly at elbows.

To create the suction and air flow, there is usually an industrial exhauster type fan that draws the dust-laden air through the system and discharges it into a cyclone collector which removes particles down to about 20 microns in size. The rest of the dust and all of the air are then discharged into the atmosphere.

With the advent of recent legislation, the cyclone collector is not efficient enough to meet the lower emission requirements. However, the cyclone collector is still an essential part of the system in most high efficiency collection systems. Although most sander dust and also some of the other dust generated in the furniture industry is smaller than 20 microns, the cyclone is an excellent precleaner for highefficiency dust collectors because it lowers the dust concentration in the air by taking out the larger particles. This enables the size of the highefficiency dust collector to be smaller.

As a matter of definition, air cleaning devices are basically divided into two separate categories: air filters and dust collectors. Air filters are used on dust concentrations of 3 grains per 1000 cubic feet or less (7000 grains being equal to 1 pound). Air filters are used on outside air for such applications as general ventilation, heating systems, and air conditioning. Dust collectors, on the other hand, handle dust concentrations from 3 grains per 1000 cubic feet to over 30 grains per cubic foot. Thus, dust collectors can handle dust concentrations many times greater than air filters and are used on industrial process air to clean it before it is exhausted, or returned to the work area.

The four basic types of dust collectors are: electrostatic precipitator, dry collector, wet collector, and fabric collector.

Generally speaking, the electrostatic collector is the most expensive and is not ideally suited for collecting wood dust.

The dry collector category includes louver, baffle, skimmer, cyclone, dynamic and impingement collectors, each of which uses centrifugal force to effect separation of the dust particles from the gas stream. For the removal of extremely dense particles, settling chambers are sometimes used. Each type is mounted over a hopper into which the collected material falls. Although dry collectors are usually used to collect dry dust, they may also be used to separate liquids from gas streams. They are termed "dry" collectors, because no liquid is added within the collector to aid in collection.

The cyclone collectors are usually designed for an inlet velocity of about 3000 to 4500 feet per minute, but the tangential velocity in the main vortex within the collector may be 10,000 feet per minute or higher. The smaller the diameter of the cyclone, the greater the centrifugal force for the same velocity in the vortex, and therefore the greater the efficiency. In addition, efficiency is dependent on inlet dust concentration and percentage of re-entrainment. Smaller diameter cyclones are used for the collection of smaller particles. To match the required capacity of the system, these smaller cyclones can be grouped together in a parallel flow arrangement (termed multiple cyclones). Where the moderately efficient large diameter cyclone, or slightly higher efficiency multiple cyclone has been acceptable in the past, the user has enjoyed low pressure drop, low initial cost and low maintenance. Today, on most woodworking applications, these cyclone collectors are simply not efficient enough to meet present standards for atmospheric exhaust.

Wet collectors, which ordinarily use water but can use other liquids, consist of air washers, packed towers, wet centrifugals, high energy venturi scrubbers, dynamic and orifice collectors. This kind of collector is moderately efficient on wood dust and is moderately priced. However, the disposal problem should be recognized. Ordinarily, the water from a wet collector is fed to a settling pond where the pollutant settles out. But, since wood dust floats, the wet collector on this application can create a water pollution problem.

The fourth category contains fabric collectors. These collectors have efficiencies over 99% if properly selected and maintained, so the exhaust from these collectors is "clean," and the need for "upgrading" later to comply with stricter air pollution control legislation is not likely. Due to the principle of their operation, they are ideally suited for the collection of wood dust, including "wood flour," most of which is smaller than 1 micron.

Fabric collectors are basically divided into two types. The envelope type, which is used for small volume flow rates of approximately 1000 to 1500 cubic feet per minute or less, is so termed because the fabric is sewn in the shape of a pocket or envelope; it is usually shaken to remove the dust from the filter medium. For volume flow rates of more than approximately 1500 cubic feet per minute, a bag or sock type collector is used, which, like the envelope type, derives its name from the shape in which the fabric is sewn. The collector may also be referred to as "bag house." Generally speaking, the larger the fabric collector, the lower the cost per cubic foot of air handled, and, therefore, the envelope type collector is rarely used except for remote locations. More frequently, the discharge from several locations are run into one large collector.

The bag or sock type collector uses either woven fabric or felt fabric. Most woven fabrics are less expensive than the felt fabrics; however, since the felt fabrics load throughout the media rather than surface loading, the felt fabric has a better initial and overall efficiency. The felt fabric is recommended particularly for those applications where it is desirable to recirculate the air back into the plant. With regard to the woven fabric, collection is obtained by building up a cake of the collected material on the dirty air side of the media. The cake provides the actual filtering or straining action which results in a high degree of removal. Since it is actually the collected material rather than the fabric which does the filtering, it is much less effective when this cake is not present, such as at initial startup and immediately after cleaning. There are numerous natural or synthetic fabrics which can be used. These bags, when used on woodworking applications, should be equipped with ground wires to bleed off the static electricity, to minimize the possibility of an explosion and/or fire.

The overall fabric collector size is determined by the fabric area required, which in turn is determined by the air-to-cloth ratio. This ratio is the cubic feet of air which can be put through one square foot of medium in one minute, or is the velocity through the medium in feet per minute, and it is a most important term when dealing with fabric collectors. It is dependent on many variables, some of which are as follows. The lower the dust concentration (expressed in grains per cubic foot), the higher the ratio can be; the larger the particle sizes of the contaminants (expressed in microns), the higher the ratio can be. Other factors which affect the ratio are the method of cleaning the bags, the desired pressure drop, and the desired maximum maintenance requirement. Of course, the maximum air-to-cloth ratio is limited by the construction of the collector and elimination of the possibility of impregnation of the medium. If the collector is too small, the pressure drop across the medium will be too high and/or the velocity through the material will be too high. This causes the dust to impregnate the filter material, with subsequent migration of the particles through the medium to discharge on the clean air side, or, the filter material can become plugged, greatly increasing the pressure drop, which reduces the air flow.

An equally important consideration is the method of cleaning the bags. There are several bag cleaning methods. One of these is a shaker which shakes or snaps the bags by use of a rotating cam arrangement. This collector ordinarily uses woven fabric to withstand the snapping action, and it will ordinarily be the largest collector for a given application due to its rather ineffective cleaning mechanism, which limits its air-to-cloth ratio to under 5:1. Also, because of its size, it is normally shipped disassembled and must be assembled at the job site. This collector is available for intermediate cleaning where the complete dust collection system can be shut down every two to eight hours in order to clean the collector, or for continuous cleaning where the collector is sectionalized into two, three, or four compartments. In the latter case, one compartment is shaken every two to eight hours. As an illustration of the continuous type of collector, assume a shaker collector divided into four compartments. One at a time, the compartments are isolated from the other three compartments for cleaning. Since all of the air must be forced through 75% of the total fabric, the pressure drop increases slightly while one compartment is being cleaned. This results in a lower suction at the hoods. However, four compartments have the advantage over three or two compartments because 75% of the fabric remains in use, whereas with three or two compartments only 66 2/3% or 50% of the material is in use at all times. To eliminate this problem, the collector can be sized so that the total amount of required fabric is available, even while one compartment is being cleaned, but when cleaning is not taking place, the collector is 100% too big if it is two compartments, 50% too big if it is three compartments, and 33 1/3% too big if it is four compartments. From this illustration, it can be seen that a four compartment continuous collector is normally the most satisfactory shaker collector, but still there may be either too little fabric when the collector is being cleaned or too much fabric when the collector is not being cleaned, depending on how the collector is sized.

Another cleaning mechanism is the pulse type of mechanism. This takes air from the plant compressed air system and discharges it into a few bags at a time, opposite to the normal air flow. This compressed air inflates the bags, breaking the dust cake on the outside of the bags and blowing the dust free of the bags, at which time it drops into a hopper at the bottom of the collector. The bags then snap back against retaining frames. This collector ordinarily uses felt fabrics rather than woven fabrics, but can use the latter as well. With this collector caution must be used to be sure that the compressed air is dry and clean—dry to eliminate excessive maintenance and clean since it operates on the clean side of the fabric. The advantages of this type of collector over the shaker collector are that it can operate at a higher air-to-cloth ratio (usually a maximum of 16:1), it is normally shipped assembled in the smaller sizes, and it operates continuously.

Another cleaning mechanism, one which has been on the market for many years, is the reverse jet mechanism. In this type of cleaner, the collector is equipped with its own source of air for cleaning, provided by a unit-mounted high pressure fan, rather than using compressed air from the plant. Spring loaded, slotted rings around the outside of each bag are mounted on a movable platform that, when activated by a pressure switch, travels up and down the bags. Meanwhile, reverse air is fed to the rings by the unit-mounted fan. This reverse air blows through the slots opposite to the normal dust-laden air flow and cleans the dust off of the inside of the bags, at which time the dust drops down into a hopper. This collector uses air-to-cloth ratios of up to 20:1. The reverse jet collector can be shipped factory assembled and can operate continuously. Felt fabrics, which enable recirculation of the air back into the plant, can be used in the reverse jet collector.

Another bag cleaning mechanism is the reverse flow unit. This gentle cleaning mechanism is ordinarily used only for high temperatures, corrosive fumes, and very delicate fabrics. The reverse flow, in addition to not being an economical selection, is not particularly suited to collecting wood dust.

When high efficiency fabric collectors are added on the discharge side of cyclones, it should be recognized that they add a pressure drop of 4 to 5 inches water gauge or higher to the system and necessitate the addition of a new fan on the clean air side of the fabric collector. This fan is usually a standard ventilating fan, sized for the same capacity as the collector and selected to develop enough pressure to overcome the resistance of the fabric collector and new ductwork. Where ductwork is installed on the discharge side of these fans, such as where the air is to be recirculated back into the building, the ductwork resistance here must also be taken into consideration.

Assuming that a plant presently has a blowpipe system installed with hoods, ductwork, fans, and cyclones, and it becomes necessary to upgrade the dust collection system, a fabric collector is the most widely accepted choice in the woodworking industry. Before calling in a consultant or a collector manufacturer, however, the owner should review the existing system to make sure it is operating at peak efficiency. It may be necessary to redesign some hoods and make some changes to the existing ductwork. As pointed out previously, the cyclones will continue to be used since they are excellent precleaners for fabric collectors, and, therefore, they too should be operating properly.

Ordinarily the plant personnel will know the capacity and dust concentration at the exit of each cyclone. However, the following should be considered in advance: the source of air and dust entering each cyclone (including which cyclones are handling sander dust); whether to recirculate the air back into the plant; whether the operation can be periodically interrupted (or whether the collector should be a continuous type); and what maintenance can reasonably be performed on this collector without overworking maintenance personnel. The air-to-cloth ratio and the recommended fabric will be determined based on these decisions.

If the plan is to recirculate the air back into the plant, the bag material should be felt, or, if the plan is to simply exhaust all of the air, woven fabric will usually be quite acceptable.

If the plant is humidified, it will be necessary to insulate the collector and ductwork exterior to the building and possibly add some heat to the air prior to entry into the dust collector. This is to eliminate condensation on the bags, which could cause the bags to become plugged and render the entire collector inoperative.

The materials of construction for the collector will probably be standard; however, if explosion venting is not standard, it should be added to the collector by the equipment manufacturer. Some collectors have reinforced steel doors which are hinged on one side and have explosion latches on the opposite side. The standard adjustable explosion latch will provide a hold-down force of approximately 5 inches water gauge operating pressure, although higher operating pressure latches can also be furnished where required. These explosion doors will open in the event of a dust explosion within the blowpipe system. Otherwise, an explosion could destroy the collector. Also, an inexpensive sprinkler system or carbon dioxide system should be added in the collector, as recommended by the owner's fire insurance company, to put out any fire within the dust collector.

The location of this equipment on the property will also have to be determined. The shaker collector, although it has the lowest initial cost, is also the largest—3 or 4 times larger than the pulse or reverse jet collector.

Regarding the cost of equipment, there are many things to consider in addition to the cost of the fabric collector, such as the new fan, motor, and V-belt drive which is required by the additional pressure drop through the collector, the ductwork cost, the installation cost, and any other accessories which must be purchased. Other considerations are what type of cleaning mechanism does the collector use, and is it intermittent or continuous? In the case of the pulse and reverse jet collectors, is it necessary to provide compressed air from the plant, or does the dust collector have its own source of air for bag cleaning? Since, periodically, maintenance people have to enter the access door on the unit to check the bags, the user should know whether he has to enter the collector on the clean air side or the dirty air side of the bags. If the outside of the bag is clean and the dust is trapped on the inside, it will be easier to install new bags and also to check for air leaks.

After the collector is installed, maintenance need not be expensive but it can be if not properly planned for in advance. The user should know the storage capacity in the hoppers, the estimated bag life, the replacement bag cost, and whether the cleaning mechanism is operated by a timer, a pressure switch, or is manual. To make maintenance easier, there are accessories available on dust collectors such as air locks, screw conveyors, ladders and platforms, trough hoppers, hopper vibrators, motor starters and control panels. Trough hoppers are highly recommended because wood dust does not flow easily, and these hoppers will facilitate flow.

Two basic factors will completely define how the collector is operating: the differential pressure and the cleanliness of the exhaust. A simple "U" tube manometer should be installed across the filter media from the dirty air side to the clean air side. By occasionally reading this manometer, a check can be kept on the filter operation. Assuming that the air is flowing at the normal design rate through the fabric medium, normal differential pressure will indicate that the cleaning mechanism, dust removal system and filter medium are properly operating. If the pressure differential reaches a point beyond the design pressure, it can mean that the collector is too small for the application. Often, however, it is a less serious problem. In some cases, a timer adjustment, pressure control adjustment, repair of the cleaning mechanism or some other minor change is needed.

The exhaust should always be clean to the eye, and visible seepage in the exhaust means that a portion of the medium is leaking. The exception to this rule is when a woven medium is brand new or after the dust cake has been broken during cleaning.

Hoppers must be regularly cleaned out; otherwise, dust will back up into the collector with subsequent damage to the bags or, possibly, to the cleaning mechanism itself. Of course, the ideal hopper is a "live bottom" hopper which uses continual dust removal by air locks or screw conveyors.

Bag life will normally be in excess of two years, and the ideal approach to bag replacement is to buy a completely new set, or, in the case of a shaker collector, at least enough for a whole compartment. Otherwise, the flow through the new medium will start at very high flow rates and may result in seepage and/or plugging of the bags. When replacement filter medium is required, it should be obtained from the original equipment manufacturer, since a substitute could lead to serious operational problems and automatically void the collector warranty.

By all means, the manufacturer's instructions on the collector, regarding regular maintenance and any work which must be performed periodically on the collector, should be studied. The importance of reading these instructions cannot be overemphasized. Also, they should be readily available to all maintenance personnel, and a maintenance schedule in accordance with these instructions should be adhered to.

In addition to the collector maintenance itself, ductwork should be periodically inspected for accumulation of dust, and any material which has settled should be removed. Fans, motors, and belts should also be regularly maintained.

With proper advance maintenance considerations, equipment selection and installation, upkeep cost should be reasonable.

GENERAL REFERENCES

Section I

AIR RESOURCE MANAGEMENT

Linsky, Benjamin. "The Relationship Between Air Pollution, Planning, and Zoning." Preprint, Air Pollution Control Assoc. Annual Meeting, Baltimore, Maryland, May 1953.

Magill, Paul L., Francis R. Holden, and Charles Ackley, Eds. *Air Pollution Handbook*, McGraw-Hill Book Company, New York, N.Y., 1956.

Faith, W. L. *Air Pollution Control.* John Wiley and Sons, Inc., 259 pp., 1959.

Meetham, A. R. *Atmospheric Pollution.* The MacMillan Company, New York, N.Y., 301 pp., 1964.

Green, H. L. and W. R. Lane. *Particulate Clouds: Dusts, Smokes and Mists.* United Press, Belfast, London, 2nd Ed. 471 pp., 1964.

Air Conservation. The Report of the Air Conservation Commission of the American Assoc. for the Advancement of Science, Publication No. 80 of the American Assoc. for the Advancement of Science, Washington, D.C. 1965.

Williams, James D. and Morman G. Edmisten. *An Air Resource Management Plan for the Nashville Metropolitan Area.* U.S. Dept. of HEW, PHS, Div. of AP, Cincinnati, Ohio, September 1965.

Wolozin, Harold, Ed. *The Economics of Air Pollution.* W. W. Norton & Company, Inc., New York, N.Y., 318 pp., 1966.

Bunyard, Francis L. and James D. Williams. *Interstate Air Pollution Study— St. Louis Area Air Pollutant Emissions Related to Actual Land Use.* U.S. Dept. HEW, PHS, NCAPC, Cincinnati, Ohio, June 1966.

Williams, James D., Frederick J. Roland, and Francis L. Bunyard. "Metropolitan Planning Program Relationships to Air Resource Management." U.S. DHEW, PHS, NCAPC, Cincinnati, Ohio, Dec. 1966.

Heller, A. N., J. S. Schueneman, and J. D. Williams. "The Air Resource Management Concept." U.S. Dept. of HEW, PHS, NCAPC, Cincinnati, Ohio, Dec. 1966.

Ridker, Ronald G. *Economic Costs of Air Pollution.* Frederick A. Praeger Pub. New York, N.Y., 214 pp., 1967.

Stern, A. C., Ed. *Air Pollution*, 2nd Ed. Vol. I Air Pollution and Its Effects, Academic Press, New York, N.Y., 1968.

Stern, A. C., Ed. *Air Pollution*, 2nd Ed. Vol. II. Academic Press, New York, N.Y., 1968.

Stern, A. C., Ed. *Air Pollution*, 2nd Ed. Vol. III Sources of Air Pollution and Their Control. Academic Press, New York, N.Y., 1968.

Rossano, A. T., Jr., Ed. *Air Pollution Control Guidebook for Management.* Environmental Science Service Div., E.R.A. Inc., Stamford, Connecticut, 214 pp., 1969.

Atkisson, Arthur and Richard S. Gaines, Ed. *Development of Air Quality Standards*. Charles E. Merrill Pub. Company, Columbus, Ohio. 220 pp., 1970.

Hagevik, George H. *Decision Making in Air Pollution Control*. Praeger Pub. New York, N.Y., 217 pp., 1970. Edelman, Sidney.

Edelman, Sidney. *The Law of Air Pollution Control*. Environmental Science Services Div., Stamford, Connecticut, 296 pp., 1970.

Strauss, Werner, Ed. *Air Pollution Control, Part I and Part II*. Wiley–Interscience, A Division of John Wiley and Sons, Inc. New York, N.Y., 1971.

A Guide for Reducing Automotive Air Pollution. Alan M. Voorhees & Associates, Inc., and Ryckman, Edgerley, Tomlinson & Associates, November 1971. NTIS No. PB 204 870.

The Relation of Land Use and Transportation Planning to Air Quality Management. Proceedings of the conference at Rutgers University, October 13, 1971. Available from Center for Urban Policy Research, Rutgers University, New Brunswick, N.J., 08903.

Noll, Kenneth E. "Transportation and Air Pollution." Journal of Traffic Engr., June 1972.

DISPERSION AND DIFFUSION

Stern, Arthur C., Ed. Symposium Proceedings on Air over Cities. U.S. Dept. of HEW, PHS Div. of Air Pollution, Sanitary Engineering Center Technical Report A62-5, Cincinnati, Ohio, November 1961.

Pasquill, F. *Atmospheric Diffusion*, Van Nostrand, New York, N.Y., 1962.

Thomas, F. W., S. B. Carpenter, F. E. Gartrell. "Stacks–How High?" Preprint, The Industrial Hygiene and Air Pollution Conference, Univ. of Texas, Austin, Texas, June 1962.

Full-Scale Study of Dispersion of Stack Gases–A Summary Report Tennessee Valley Authority and Public Health Service, Chattanooga, Tennessee, August 1964.

May, James W. *The Physics of Air*, 5th Ed., American Air Filter Company, Inc. Louisville, Ky., 1964.

Central Electricity Research Laboratory Symposium on Chimney Plume Rise and Dispersion. Atmospheric Environment, Pergamon Press, Vol. 1, p. 351–440, 1967.

Hage, K. D., P. S. Brown, G. Arnason, S. Lozorick, and M. Levitz. *A Computer Program for the Fall and Dispersion of Particles in the Atmosphere*. Clearinghouse for Federal Scientific and Technical Information, Springfield, Va., April 1967.

Tall Stacks, Various Atmospheric Phenomena, and Related Aspects. U.S. Dept. of HEW, PHS, National Air Pollution Control Administration Pub. No. APTD 69-12, Arlington, Va., May 1969.

The Tall Stack for Air Pollution Control on Large Fossil-Fueled Power Plants. A Collection of Recent Papers with an Introduction by Philip Sporn, American Electric Power Co., New York, N.Y., May 1967.

"Giant Stack Will Vent Sulfur Oxides Above Smog Ceiling." Chem. Eng., 74(17):104, Aug. 14, 1967.

Duprey, R. L. *Compilation of Air Pollutant Emission Factors.* Public Health Service, Durham, N.C., National Center for Air Pollution Control, Publication No. 999-AP-42, 67 p., 1968. 126 refs.

Lowry, William P. and Richard W. Boubel. *Meteorological Concepts in Air Sanitation*, 2nd Ed., Corvallis, Oregon, 1968.

Frankenberg, T. T. "High Stacks for the Diffusion of Sulfur Dioxide and Other Gases Emitted by Electric Power Plants." Am. Ind. Hyg. Assoc. J., 29(2): 181–185, March–April 1968. 6 refs.

Slade, David H., Ed. *Meteorology and Atomic Energy.* United States Atomic Energy Commission, Div. of Technical Information, TID-24190, July 1968.

Full-Scale Study of Plume Rise at Large Electric Generating Stations, Report by Tennessee Valley Authority, Muscle Shoals, Alabama, September 1968.

Niemeyer, L. E. and F. A. Schiermeier. "The Tall Stack: A Question of Effectiveness in Air Pollution Management." Preprint, National Air Pollution Control Administration, Cincinnati, Ohio, Air Resources Lab., 15 p., Nov. 1968.

Briggs, G. A. *Plume Rise.* A & C Critical Review Series, TID-25075, Clearinghouse for Federal Scientific and Technical Information, Springfield, Va., 1969.

Turner, Bruce D. *Workbook of Atmospheric Dispersion Estimates.* U.S. Dept. of HEW, PHS, National Air Pollution Control Admin. Pub. No. 999-AP-26, Cincinnati, Ohio, 1969.

"A Discussion on Recent Research in Air Pollution." Philosophical Transaction of the Royal Society of London, Vol. 265, p. 139–318, No. 161, Nov. 13, 1969.

Stern, Arthur C., Ed. Proceedings of Symposium on Multiple-Source Urban Diffusion Models, U.S. Environmental Protection Agency, Office of Air Programs Pub. No. AP-86, Research Triangle Park, N.C., 1970.

Report on Full-Scale Study of Inversion Breakup at Large Power Plants. Tennessee Valley Authority, Muscle Shoals, Alabama, March 1970.

Holzworth, George C. *Mixing Heights, Wind Speeds, and Potential for Urban Air Pollution Throughout the Contiguous United States.* U.S. Environmental Protection Agency, Office of Air Programs Pub. No. AP-101, Research Traingle Park, N.C., January 1972.

AIR POLLUTION STANDARDS

Yocom, John E. "Air Pollution Regulations—Their Growing Impact on Engineering Decisions." Chem. Eng., 29(15):103–114, July 23, 1962.

Stern, Arthur C. "Air Pollution Standards." In: Air Pollution. Arthur C. Stern (ed.), Vol. 3, 2nd ed., New York, Academic Press, 1968, Chapt. 51, p. 601–718. 22 refs.

Martin, Robert and Lloyd Symington. "A Guide to the Air Quality Act of 1967." Law and Contemporary Problems. 33(2):239–274, Spring 1968.

"Recommended Air Quality Standards Based on Health Effects" Clean Air Quarterly, 12(41):5–29, December 1968.

Hershaft, A. Air Pollution Control: A Critical Overview. Grumman Aircraft Engineering Corp., Bethpage, N.Y., Research Dept., RM-446, 28 p. June 1969.

McGuire, Terry and Kenneth E. Noll. "Relationships between Concentrations of Atmospheric Pollutants and Averaging Times." Preprint, California State Dept. of Public Health, Berkeley; Calif. Univ., Berkeley, Public Health Service, Washington, D.C., National Air Pollution Control Admin. 19 p., 1970.

Farmer, Jack R., Philip J. Breibaum, and Joseph A. Tikvart. "Proceedings from Air Quality Standards to Emission Standards. Preprint, Air Pollution Control Association, New York City, 83 p., 1970.

The Clean Air Act. United States 92nd Congress. 56 p., Dec. 1970.

Heuss, Jon M., George M. Nebel, and Joseph M. Colucci. "National Air Quality Standards for Automotive Pollutants: A Critical Review." Preprint, Air Pollution Control Assoc., Pittsburgh, Pa., 24 p., 1971. 52 refs.

Potter, Allen E. "Eight Steps to a Successful Control Program." Foundry, 99(4):AP26–AP29, April 1971.

Ellis, Howard M. and Simon K. Mencher. "An Evaluation of the Proposed National Ambient Air Quality Standards for Particulate Matter, Sulfur Dioxide, and Nitrogen Dioxide." J. Air Pollution Control Assoc., 21(6): 348–351, June 1971. 30 refs.

Welby, Paul. "Measuring up to Air Pollution Control." Plant Eng., 25(15): 66–67, July 22, 1971.

Leithe, W. The Analaysis of Air Pollutants (Ann Arbor, Michigan: Ann Arbor Science Publishers, 1971).

Hesketh, Howard E. Understanding and Controlling Air Pollution (Ann Arbor, Michigan: Ann Arbor Science Publishers, 1972).

Patterson, D. J., and N. A. Henein. Emissions from Combustion Engines and Their Control (Ann Arbor, Michigan: Ann Arbor Science Publishers, 1972).

Brenchley, David L., and C. David Turley. Industrial Source Sampling (Ann Arbor, Michigan: Ann Arbor Science Publishers, 1973).

McCrone, Walter C., and John G. Delly. The Particle Atlas, Edition Two, 4 vols. (Ann Arbor, Michigan: Ann Arbor Science Publishers, 1973).

AIR MONITORING AND SOURCE SAMPLING

Brown, E. L. "Flue Gas Dust Sampling." In: Gas Purification Processes. G. Nonhebel (ed.), London, George Newnes Ltd., 1964, Chapt. 14, Part B p. 601–621. 2 refs.

Devarkin, Howard, Robert L. Chass, Albert P. Fudwich, Carl V. Kanter. *Air Pollution Source Testing Manual*. Air Pollution Control District, Los Angeles County, Calif. November 1965.

Selected Methods for the Measurement of Air Pollutants. U.S. Dept. of HEW, PHS, Division of Air Pollution Pub. No. 999-AP-11, Cincinnati, Ohio, May 1965.

Haaland, Harold H., Ed. *Methods for Determination of Velocity, Volume, Dust, and Mist Content of Gases*. 7th Ed. Western Precipitation Division, Joy Manufacturing Company, Los Angeles, Calif., 1968.

Duncan, Joseph R. and Auburn E. Owen, Jr. "An Exercise in Source Sampling Methods." Preprint, Tennessee Manufacturer's Association Workshop, Nashville, Tennessee, March 1970.

Duncan, Joseph R. "The Application of Telemetry Systems for Air Quality Monitoring at Large Power Plants." Proceedings of the Ninth Annual Air Pollution Control Conf., Purdue Univ., Lafayette, Indiana, October 13, 1970.

Cooper, H. B. H., Jr., and A. T. Rossano, Jr. *Source Testing for Air Pollution Control*. Environmental Science Services Div., Wilton, Connecticut, 1971.

Serper, Allen. "A Look at the Methods for Measuring Air Pollutants." Eng. Mining J., 172(4):124–28, April 1971.

Martin, Robert M. *Construction Details of Isokinetic Source-Sampling Equipment*. U.S. Environmental Protection Agency, Air Pollution Control Office, Pub. No. APTD-0581, Research Triangle Park, N.C., April, 1971.

Nelson, Gary O. *Controlled Test Atmospheres: Principles and Techniques* (Ann Arbor, Michigan: Ann Arbor Science Publishers, 1971).

Brenchley, David L., and C. David Turley. *Industrial Source Sampling* (Ann Arbor, Michigan: Ann Arbor Science Publishers, 1973).

URBAN PLANNING

Rydell, C. Peter and Gretchen Schwarz. "Air Pollution and Urban Form: A Review of Current Literature." American Institute of Planners Journal, V. 34(2), 1968.

Air Quality Display Model, Developed by TRW Systems Group under Contract No. PH 22-68-60, U.S. Dept. of HEW, PHS, National Air Pollution Control Admin. November 1969.

Air Quality Implementation Planning Program, Prepared by TRW Systems
    Group, Washington Operations under Contract No. PH 22-68-60, U.S.
    Environmental Protection Agency, National Air Pollution Control
    Admin., Washington, D.C., November 1970.

Yocom, John E., George F. Collins, and Norman E. Bowne. "Plant Site
    Selection." Chem. Eng., 78(14):164–168. June 12, 1971.

GENERAL REFERENCES

Section II

COMBUSTION

Johnstone, H. F. "The Elimination of Sulphur Compounds from Boiler Furnace Gases. Part I." Steam Eng., 1932:153–154, Jan. 1932. 5 refs. Part II. Ibid, 1932:208–211, Feb. 1932. 1 ref.

White, H. J. "Effect of Flyash Characteristics on Collector Performance." Air Repair 5 (1), 37–50, 62 (May 1955.)

Stairmand, C. J. and R. M. Kelsey. "The Role of the Cyclone in Reducing Atmospheric Pollution." Chem. Ind.(London), 1955:1324–1330, Oct. 15, 1955. 5 refs.

Monkhouse, A. C. and H. E. Newall. "Industrial Gases–Recovery of Sulphur Dioxide." Society of Chemical Industry, London (England), Disposal Ind. Waste Mat. Conf. Sheffield, England, 1956, p. 103–107. 17 refs. (April 17–19).

Hopps, George L., A. A. Berk, and J. F. Barkley. "Tests of Additives to Control Soot Deposition in Oil-Fired Boilers." Bureau of Mines, Washington, D.C., Dept. of Investigations 5947, 19 p., 1962. 15 refs.

Sensenbaugh, J. D. "Formation and Control of Sulfur Oxides in Boilers." J. Air Pollution Control Assoc., 12(12):567–569, 591, Dec. 1962. 32 refs.

Stewart, I. McC. "Solid Fuel Firing of Small Industrial Boilers in the 'Clean Air' Age." Proc. Clean Air Conf., Univ. New South Wales, 1962, Paper 23, Vol. 2, 16 p.

"Coal-Fired Heating Plant Package: Phase II Report." Pope, Evans and Robbins, New York. OCR Contract 14-01-0001-242, 63 p., Nov. 1, 1963. CF-STI:PB 181–585.

Wangerin, D. D. "Waste Heat Boilers–Principles and Applications." Proc. Am. Power Conf. (Presented at the 26th Annual Meeting, American Power Conference, Chicago, Ill., Apr. 14–16, 1964) 26, 682–91, Apr. 1964.

Mayer, M. "A Compilation of Air Pollutant Emission Factors for Combustion Processes, Gasoline Evaporation, and Selected Industrial Processes." Public Health Service, Cincinnati, Ohio, Div. of Air Pollution, May 1965, 53 p.

"Air Pollution Control: The Sulfur Problem." Coal Age, 70(8):58–62, Aug. 1965.

"Practical Applications of Additives to Control Air Pollution–For Use with Petroleum Fuels." National Petroleum Refiners Association, Washington, D.C., FL-66-46(a), 318 p., 1966. 6 refs.

317

Voelker, E. M. "Control of Air Pollution from Industrial and Household Incinerators." Proc. Natl. Conf. Air Pollution, 3rd, Washington, D. C., 1966, pp. 332–8.

Frankel, J. I. "Incineration of Process Wastes." Chem. Eng. 73(18):91–96, Aug. 29, 1966.

Hescheles, C. A. "Industrial Waste Analysis and Boiler Performance Test Burning Wastes." Preprint. (Presented at the Winter Annual Meeting and Energy Systems Exposition, American Society of Mechanical Engineers, New York City, Nov. 27–Dec. 1, 1966.)

Challis, J. A. "Three Industrial Incinerator Problems." Proc. Natl. Incinerator Conf. New York, 1966 208–18, 1966.

"Control of Industrial Boilers by Oxygen Analysis of Flue Gases." Power Works Eng. 61 (723), 57–61 (Sept. 1966).

Fernandes, J. H., J. D. Sensenbaugh, and D. G. Peterson. "Boiler Emissions and Their Control." Combustion Engineering, Inc., Windsor, Conn., and Air Preheater Co., Wellsville, N.Y.

Burdock, J. L. "Fly Ash Collection from Oil-Fired Boilers." Preprint, UOP Air Correction Div., Greenwich, Conn., 15 p., 1966. 4 refs.

Hescheles, C. A. "Burning Industrial Wastes." Proc. Mecar Symp., Incineration of Solid Wastes, New York City, 1967. pp. 60–74.

Byers, R. E. "Combustion Airflow: Its Measurement and Control." TAPPI, 50(4):52A–58A, April 1967. 8 refs.

Walker, A. B. "Emission Characteristics from Industrial Boilers." Air Eng., 9(8):17–19, Aug. 1967.

Lock, A. E. "Reduction of Atmospheric Pollution by Efficient Combustion Control." Plant Eng. (London), 11(5):305–309, May 1967.

Perry, Harry and J. H. Field. "Coal and Sulfur Dioxide Pollution." American Society of Mechanical Engineers, New York, Paper 67. WA/PID-6 9 p., 1967. 19 refs.

Polglase, William L. "Boilers Used as Afterburners." In: Air Pollution Engineering Manual. (Air Pollution Cont. Dist., Co. of Los Angeles.) John A. Danielson (comp. and ed.), Public Health Service, Cincinnati, Ohio, Natl. Center for Air Poll. Cont., PHS-Pub-999-AP-40, p. 187–192, 1967. GPO:806-614-30.

Fernandes, J. H. "Incinerator Air Pollution Control." American Society of Mechanical Engineers, New York, Incinerator Div., Proc. Natl. Incinerator Conf., New York, 1968, p. 101–116. 44 refs. (May 5–8).

Zabroske, Tony A. "Boiler Conversion Reduces Costs and Air Pollution." Plant Eng., 22(6):96, 98, March 21, 1968.

Rutz, P. "Boiler Plants for Burning Industrial Wastes." Sulzer Tech. Rev. (Switz.) 3:99–108, 1968.

Green, Bobby L. "Boiler for Bark-Burning." Power Eng., 72(9):52–53, Sept. 1968.

Tada, Mitsuru. "Industrial Waste Incineration by Fluidizing System." (Sangyo haikibutsu no ryudoshokyakuho). Text in Japanese. Kogai to Taisaku (J. Pollution Control), 5(7):529–533, July 1969;

Borio, Richard W., Robert P. Hensel, Richard C. Ulmer, Hilary A. Grabowski, Edwin B. Wilson, and Joseph W. Leonard. "The Control of High-Temperature Fire-Side Corrosion in Utility Coal-Fired Boilers." Combustion Engineering, Inc., Windsor, Conn., Research and Product Development, Contract 14-01-0001-485, OCR R&D Dept. 41, 224 p., April 25, 1969, 35 refs.

Douglas, Jack. "Instruments and Controls for Industrial Power Plants." Nat. Eng., 73(7):10–12, July 1969.

"A New Approach to Industrial Air Pollution Control." Heating Ventilating Engr. J. Air Conditioning (London), 44(517):80–85, Aug. 1970.

Spaite, Paul W. and Robert P. Hangebrauck. "Pollution from Combustion of Fossil Fuels." In: Air Pollution-1970 Part I. 91st Congress (Senate), Second Session on S.3229, S.3466, S.3546, p. 172–181, 1970. 3 refs.

Feldkircher, James L. "Rebirth of a Boilerhouse." Preprint, American Society of Mechanical Engineers, New York, 5 p., 1970.

Ehrenfeld, John R., Josette C. Goldish, Ronald Orner, and Ralph H. Bernstein. "Pollution from Stationary Fossil-Fuel Burning Combustion Equipment to 1990. A Systems Study of Emissions and Control." Preprint, International Union of Air Pollution Prevention Associations, 32 p., 1970. 18 refs.

"Nationwide Inventory of Air Pollutant Emissions." National Air Pollution Control Administration, Raleigh, N.C., Div. of Air Quality and Emission Data. Pub. AP-73, 36 p., Aug. 1970. 13 refs. NTIS:PB 196304.

Coates, N. H., P. S. Lewis, and J. W. Eckerd. "Combustion of Coal in Fluidized Beds." Trans. AIME (Am. Inst. Mining Metallurgical and Petroleum Engrs.), 247(3):208–210, Sept. 1970.

Spaite, Paul W. and Robert P. Hangebrauck. "HEW Spells Out Air-Quality Goals." Elec. World, 173(20):25–27, May 18, 1970.

Hall, R. E., J. H. Wasser, and E. E. Berkau. "NAPCA Combustion Research Programs to Control Pollutant Emissions from Domestic and Commercial Heating Systems." Preprint, National Oil Fuel Inst., New York, 18 p., 1970.

Govan, Francis A. "Control Equipment Not Always the Answer to Pollution Control." Bldg. Systems Design, 68(2):16, 17, 37, Feb. 1971.

Sticksel, Philip R. and Richard B. Engdahl. "Derivation of the Emission Data and Projections Used in Planning." In: The Federal R and D Plan for Air Pollution Control by Process Modification. Battelle Memorial Inst., Columbus, Ohio, Columbus Labs., APCO Contract CPA 22-69-147, Rept. APTD-0643, p. B-1 to B-19, Jan. 11, 1971. 11 refs. NTIS:PB 198066.

Pinheiro, George. "Precipitators for Oil-Fired Boilers." Power Eng., 75(4):52–54, April 1971.

Papamarcos, John. "Fuel Oil Additive Passes Tests." Power Eng., 75(4):46–48, April 1971. 1 ref.

Frazier, J. F. "Removal of Sulfur Oxides from Industrial Boiler Flue Gases." Natl. Eng., 75(8):6–9, Aug. 1971. 4 refs.

Walsh, W. H. and J. A. Waddell. "Obtaining and Maintaining Low Excess Air Operation on an Industrial Boiler." Preprint, Illinois Inst. of Tech., Chicago, Technical Center, 8 p., 1971.

Barrett, R. E. and D. W. Locklin. "Industrial Steam Generation and Commercial and Residential Heating." In: The Federal R and D Plan for Air-Pollution Control by Combustion-Process Modification. Battelle Memorial Inst., Columbus, Ohio, Columbus Labs., 1971, APCO Contract CPA 22-69-147, Rept. APTD-0643, p. V-1 to V-47, Jan. 11, 1971. 48 refs. NTIS: PB 198066.

Barron, A. V., Jr. "Particulate and SO_2 Control Technology for the Small and Medium Coal-Fired Boiler." Combustion, 43(4):44–56, Oct. 1971.

Plumley, A. L. "Fossil Fuel and the Environment–Present Systems and Their Emissions." Combustion, 43(4):36–43, Oct. 1971. 21 refs.

Baddams, H. W. "Industrial Combustion of Oil Fuels." Clean Air (J. Clean Air Soc. Australia New Zealand), 5(2):31–37, May 1971. 6 refs.

POWER GENERATION

Flodin, C. R. and H. H. Haaland. "Some Factors Affecting Fly-Ash Collector Performance on Large Pulverized Fuel-Fired Boilers." Air Repair 5 (1), 27–32 (May 1955).

Bienstock, D., J. H. Field, and H. E. Benson. "Sulfur Dioxide in Atmospheric Pollution, and Methods of Control." (Proc. Symp. Atmospheric Chemistry of Chlorine and Sulfur Compounds, Cincinnati, Ohio), 1957. (1959). pp. 54–62. (Geophysical Monograph No. 3.)

Kirkwood, J. B. "Electrostatic Precipitators for the Collection of Fly Ash from Large Pulverised Fuel Fired Boilers." Proc. Clean Air Conf., Univ. New South Wales, 1962, Paper 14. Vol. 2, 20 p.

Stone, G. N. and A. J. Clarke. "Power Stations and Clean Air." Central Electricity Generating Board. (England) 1963. 12 pp.

Rees, R. L. "Removal of Sulfur Dioxide from Power-Plant Stack Gases." In: Problems and Control of Air Pollution. F. S. Mallette (ed.), New York, Reinhold, 1955, Chapt. 14, p. 143–154. 16 refs.

Billinge, B. H. M., A. C. Collins, B. Hearn, and H. G. Masterson. "The Dry Removal of Sulphur Dioxide from Flue Gases." Cost Estimates. Preprint. 1963.

Bienstock, D., J. H. Field, and J. G. Myers. "Removal of Sulfur Oxides from Flue Gas with Alkalized Alumina at Elevated Temperatures." J. Eng. Power 86, (3) 353-60, July 1964.

Duzy, A. F. "American Coal Characteristics and Their Effects on the Design of Steam Generating Units." Preprint, American Society of Mechanical Engineers, New York, 8 p., 1959. 13 refs.

"Steam-Electric Plant Factors, 1960." Fuel consumption and costs, plant capacity, net generation, 1960, and programmed capacity, 1961–64. Washington, D.C., National Coal Association, 1961, 37 p.

Cuffe, S. T., R. W. Gerstle, A. A. Orning, and C. H. Schwartz. "Air Pollutant Emissions from Coal-Fired Power Plants; Report No. 1." J. Air Pollution Control Assoc. 14, (9) 353-62, Sept. 1964.

Frankenberg, T. T. "Removal of Sulfur from Products of Combustion," Proc. Am. Petrol Industry. 45, (3) 365-70, 1965.

Germerdonk, R. "Scrubbing Sulfur Dioxide from Flue Gases." Auswaschen von Schwefeldioxyd aus Rauchgasen. Chem. Ingr. Tech. 37, (11) 1136-9, Nov. 1965. Ger.

Craxford, S. R. "Air Pollution from Power Stations." Smokeless Air (London), Vol. 36:123–128, 1965.

Rohrman, F. A., J. H. Ludwig, and B. J. Steigerwald. "Coal Utilization and Atmospheric Pollution." Coal–Wherever Coal is Concerned 19, (4) 5–7, Apr. 1965.

Kata, J. "The Effective Collection of Fly Ash at Pulverized Coal-Fired Plants." J. Air Pollution Control Assoc. Vol. 15 (11):525–528, Nov. 1965.

Diehl, E. K. and E. A. Zawadzki. "Contaminants in Flue Gases–And Methods for Removal." Coal Age, Vol. 70:70–74, Dec. 1965.

Gerstle, R. W., S. T. Cuffe, A. A. Orning, and C. H. Schwartz. "Air Pollutant Emissions from Coal-Fired Power Plants, Report No. 2." J. Air Pollution Control Assoc. 15, (2) 59–64, Feb. 1965.

Jones, J. R. "Contribution of the Coal Industry to Solving the Problem of Air Pollution Control." Proc. Am. Power Conf. 27, 126–36, Apr. 1965.

McKelvey, V. E. and D. C. Duncan. "United States and World Resources of Energy." Am. Chem. Soc., Pittsburgh, Pa., Div. Fuel Chem., Preprints, (2):1–17, 1965.

Benson, H. E. and C. L. Tsaros. "Conversion of Fossil Fuels to Utility Gas." Am. Chem. Soc., Pittsburgh, Pa., Div. Fuel Chem., Preprints, 9 (2): 104–113, 1965. 9 refs.

Schlesinger, M. D., G. U. Dinneen, S. Katell. "Conversion of Fossil Fuels to Liquid Fuels." Am. Chem. Soc., Pittsburg, Pa., Div. Fuel Chem., Preprints, 9 (2):120–126, 1965. 12 refs.

Clendenin, J. D. "The Utilization of Coal." Am. Chem. Soc., Pittsburgh, Pa., Div. Fuel Chem. Preprints, 9 (2):222, 1965.

Morrison, Warren E. "The Energy Dilemma–Which Fuel, What Market, When." Preprint No. 65K302, Society of Mining Engineers, AIME, N.Y., 26 p., 1965.

Kata, J. "The Effective Collection of Fly Ash at Pulverized Coal-Fired Plants." J. Air Pollution Control Association. Vol. 15 (11):525–528, Nov. 1965.

Diehl, E. K. and E. A. Zawadzki. "Contaminants in Flue Gases—And Methods for Removal." Coal Age, Vol. 70:70–74, Dec. 1965.

Smith, W. S. and C. W. Gruber. "Atmospheric Emissions from Coal Combustion—An Inventory Guide." Public Health Service, Cincinnati, Ohio, Division of Air Pollution. (999-AP-24.) Apr. 1966. 117 pp.

Rohrman, F. A., B. J. Steigerwald, and J. H. Ludwig. "The Role of the Power Plant in Sulfur Dioxide Emissions: 1940–2000." Preprint. 1966.

Garvey, J. R. "Air Pollution and the Coal Industry." Mining Congr. J. pp. 55–65. Aug. 1966.

"Promise Seen in Stack-Gas SO_2 Removal." Oil Gas J., p. 53, May 2, 1966.

Perry, H. "Potential for Reduction of Sulfur in Coal by Other than Conventional Cleaning Methods." Preprint. (Presented at the Symposium on Economics of Air Pollution Control, 59th National Meeting of the American Inst. of Chemical Engineers, Columbus, Ohio, May 15–18, 1966, Paper No. 24 E.)

Engelbrecht, H. L. "Electrostatic Precipitators in Thermal Power Stations Using Low Grade Coal." Preprint. (Presented at the 28th Annual Meeting, American Power Conference, April 26–28, 1966.)

Katell, Sidney. "An Evaluation of Dry Processes for the Removal of Sulfur Dioxide from Power-Plant Flue Gases." Preprint, Bureau of Mines, Morgantown, W. Va., Morgantown Coal Research Center, 23 p., 1966. 8 refs.

Katell, S. "Removing Sulfur Dioxide from Flue Gases." Chem. Eng. Progr. 62, (10) 67–73, Oct. 1966.

Gartrell, F. E. "Control of Air Pollution from Large Thermal Power Stations." Rev. Soc. Roy. Belge Ingrs. Ind. (Brussels) (11) 471–82, Nov. 1966.

Gosselin, A. E., Jr., and L. W. Lemon. "Bag Filterhouse Pilot Installation on a Coal-Fired Boiler—Preliminary Report and Objectives." Proc. Am. Power Conf., Vol. 28, p. 534–545, 1966. 3 refs.

Williamson, Gerald V. and John F. McLaughlin. "Air Pollution, Its Relation to the Expanding Power Industry." Union Electric Co., St. Louis, Mo., 16 p., 1966. 6 refs.

"SO_2 Stack Gas Gives (NH_4) $2SO_4$." Chem. Eng. News, 44(26):23, June 27, 1966.

Watkins, E. R. and K. Darby. "Electrostatic Precipitation for Large Boilers." Proc. Inst. Mech. Engrs. (London), vol. 181, part 3N:78–89, 1966–1967. 3 refs.

Shale, C. C. "Progress in High-Temperature Electrostatic Precipitation." J. Air Pollution Control Assoc. 17, (3) 159–60, Mar. 1967.

Pollock, W. A., G. Frieling, and J. P. Tomany. "Sulfur Dioxide and Fly Ash Removal from Coal Burning Power Plants." Air Eng. 9(9), 24–8 (Sept. 1967).

Atsukawa, M., Y. Nishimoto, and K. Matsumoto. "Dry Process SO_2 Removal Method." Tech. Rev. Mitsubishi Heavy Ind. (Tokyo) 4, (1) 33–8, 1967.

"Report on Sulfur Dioxide and Fly Ash Emissions from Electric Utility Boilers." Public Service Electric and Gas Co., Trenton, N.J.; Jersey Central Power & Light Co./New Jersey Power & Light Co., Morristown. (February 24, 1967). 67 pp.

Teller, Aaron J. "Recovery of Sulfur Oxides from Stack Gases." Proc. MECAR Symp., New Developments in Air Pollution Control, Metropolitan Engineers Council on Air Resources, New York City, p. 1–11, Oct. 23, 1967.

Slack, A. V. "Removal of Sulfur Oxides from Power Plant Stack Gases: Outline of Major Problems." Proc. MECAR Symp., New Developments in Air Pollution Control, Metropolitan Engineers Council on Air Resources, New York City, p. 42–49, Oct. 23, 1967. 2 refs.

Haynes, W. P. "Current Work at the Bureau of Mines on Recovery of Sulfur Oxides from Stack Gas." Proc. MECAR Symp., New Developments in Air Pollution Control, Metropolitan Engineers Council on Air Resources, New York City, p. 50–61, Oct. 23, 1967.

Cahill, William J., Jr. "Control of Particulate Emissions on Electric Utilities Boilers." Proc. MECAR Symp., New Developments in Air Pollution Control, Metropolitan Engineers Council on Air Resources, New York City, p. 74–84, Oct. 23, 1967.

Squires, Arthur M. "Air Pollution: The Control of SO_2 from Power Stacks. Part IV Power Generation with Clean Fuels." Chem. Eng., 74 (26):101–109, Dec. 18, 1967. 35 refs.

Squires, Arthur M. "Air Pollution: The Control of SO_2 from Power Stacks. Part I–The Removal of Sulfur from Fuels." Chem. Eng., 74 (23):260–268, Nov. 6, 1967. 35 refs.

Squires, Arthur M. "Air Pollution: The Control of SO_2 from Power Stacks. Part II. The Removal of SO_2 from Stack Gases." Chem. Eng., 74 (24): 133–140, Nov. 20, 1967. 34 refs.

Ito, Akio, Tadao Shirasawa, Tomio Ohyanagi, and Yukio Tamori. "Packed Coal Bed as a Dust Collector (II)." Taiki Osen Kenkyu (J. Japan Soc. Air Pollution), 2 (1):98–100, 1967. Translated from Japanese. 8 p.

Slack, A. V. "Air Pollution: The Control of SO_2 from Power Stacks. Part III–Processes for Recovering SO_2 " Chem. Eng., Vol. 74, p. 188–196, Dec. 4, 1967. 4 refs.

Gerstle, R. W. "Estimating Particulate and Sulfur Dioxide Emissions from Fuel Combustion." Preprint, Public Health Service, Cincinnati, Ohio, National Air Pollution Control Administration, 12 p., Aug. 1968. 4 refs.

Ritchings, F. A. "Raw Energy Sources for Electric Generation." IEEE (Inst. Elec. Electron, Engrs) Spectrum, 5 (8):34–45, Aug. 1968.

Perry, H., J. McGee, and D. Strimbeck. "Electricity from Coal. The Cycles. Part 2." Mech. Eng., 90 (12):44–47, Dec. 1968.

Ochs, Hans-Joachim. "Dust and Gaseous Emissions from Power Plants." (Staub-und gasfoermige Emissionen von Kraftwerken). Text in German. Wasser Luft Betrieb, 12 (5):284–288, May 1968. 3 refs.

Potter, A. E., R. E. Harrington, and P. W. Spaite. "Limestone-Dolomite Processes for Flue Gas Desulfurization." Air Eng., 10 (4):22–27, April 1968. 17 refs.

Reese, J. T. and Joseph Greco. "Experience with Electrostatic Fly-Ash Collection Equipment Serving Steam-Electric Generating Plants." J. Air Pollution Control Assoc., 18 (8):523–528, Aug. 1968. 8 refs.

Crawford, W. D. "The Cost of Clean Energy." Preprint, Consolidated Edison Co. of New York, Inc., 14 p., 1968.

"Sulfur Oxide Removal from Power Plant Stack Gas: Conceptual Design and Cost Study. Sorption by Limestone or Lime: Dry Process." Tennessee Valley Authority. 91 p., 1968. 57 refs. CFSTI:PB-178971.

McLaughlin, J. F., Jr. "Progress in Meeting Power Plant Air Pollution Problems." EEI Bull. 36 (5), 155–9 (May 1968).

"Monsanto Process Removes Sulfur Dioxide from Gases." Edison Elec. Inst. Bull., 36 (9):320, Oct. 1968.

Falkenberry, H. L. and A. V. Slack. "Removal of SO_2 from Power Plant Stack Gases by Limestone Injection." Preprint, Tennessee Valley Authority, Chattanooga, (35) p., 1968. (8) refs.

"Development of a Molten Carbonate Process for Removal of Sulfur Dioxide from Power Plant Stack Gases. (Summary Report.)" North American Rockwell Corp., Canoga Park, Calif., Atomics International Div. Contract PH 86-67-128, AI-68-104, 155 p., 1968. 8 refs. CFSTI:PB 179908.

Baxter, W. A. "Recent Electrostatic Precipitator Experience with Ammonia Conditioning of Power Boiler Flue Gases." J. Air Pollution Assoc., 18 (12):817–820, Dec. 1968. 9 ref.

"SO_2 Control Processes for Stack Gases Reach Commercial Status." Environ. Sci. Technol., 2 (11):994–997, Nov. 1968.

Thring, M. W. "Fuel and Energy Problems in the Next Thirty Years." Preprint, Combustion Engineering Assoc. Hayes, Middx., Great Britain, 21 p., 1969.

Zawadzki, Edward A. "Status of the Development of Processes for Controlling SO_2 Emissions from Stationary Sources." Preprint, National Limestone Inst., Inc., Washington, D.C., 15 p., 1969.

"Sulfur Oxide Removal from Power Plant Stack Gas. Use of Limestone in Wet-Scrubbing Process." Tennessee Valley Authority, 104 p., 1969. 22 refs. CFSTI:PB 183908.

Shale, C. C. and G. E. Fasching. "Operating Characteristics of a High-Temperature Electrostatic Precipitator." Bureau of Mines, Morgantown, W. Va., Morgantown Coal Research Center, RI 7276, 19 p., July 1969. 19 refs. CFSTI:PB 185549.

Lower, H. J. "Reduction of Emission of Pollutants. Recent Advances in Electrostatic Precipitators for Dust Removal." Phil. Trans. Roy. Soc. London, Ser. A, 265 (1161):301–307, Nov. 13, 1969. 16 refs.

Stankus, L. "NAPCA's Search for Flue Gas Desulfurizing Processes." Preprint, 22 p., 1969. (Presented at the Gordon Research Conference, Aug. 18–22, 1969.)

Gambs, Gerard C. "The Electric Utility Industry: Future Fuel Requirements 1970–1990." Mech. Eng., 92 (4):42–48, April 1970.

Spaite, Paul W. and Robert P. Hangebrauck. "Pollution from Combustion of Fossil Fuels." In: Air Pollution-1970 Part I. 91st Congress (Senate), Second Session on S.3229, S.3466, S.3546, p. 172–181, 1970. 3 refs.

Perrine, Richard L. and Limin Hsueh. "Power and Industry: Particulates." In: Project Clean Air. California Univ., Berkeley, Task Force 5, Vol. 1, Section 11, 3 p., Sept. 1, 1970.

Hangebrauck, Robert P. and Paul W. Spaite. "Pollution from Power Production." Preprint, National Limestone Inst. Inc., 21 p., Jan. 1970. 9 refs.

Craig, T. L. "Recovery of Sulfur Dioxide from Stack Gases: The Wellman-Lord SO_2 Recovery Process." Preprint, Kentucky Univ., Lexington, 11 p., 1970.

Martin, J. R., W. C. Taylor, and A. L. Plumley. "The C-E Air Pollution Control System." Nat. Eng., 74 (6):8–12, June 1970. C-E's Air Pollution Control System. Ibid., 74 (7):8–10, July 1970. 7 refs.

Henke, William G. "The New 'Hot' Electrostatic Precipitator." Combustion, 42 (4):50–53, Oct. 1970.

Harrison, D. and A. Saleem. "Where We Stand in Sulphur Dioxide Control." Mod. Power Eng., 64 (6):62–63, June 1970.

Reid, William T. "What About Air Pollution by Power Plants." Battelle Res. Outlook, 2 (3):21–24, 1970.

GENERAL REFERENCES

Section III

Coulter, R. S. "Smoke, Dust, Fumes Closely Controlled in Electric Furnaces." Iron Age, 173(2):107–110, Jan. 14, 1954.

Shaffer, N. R. and M. A. Brower. "Air Pollution: Furnace Types and Sizes Dictate Most Effective Controls. Part I." Iron Age, 175(17):100–102, April 28, 1955. Part II. Ibid, 175(18):110–112, May 5, 1955. Part III, Ibid, 175(19):100–102, May 12, 1955.

Basse, B. "Gases Cleaned by the Use of Scrubbers." Blast Furnace Steel Plant 1307–12, Nov. 1956.

Silverman, L. "Research and Development of Equipment for Cleaning of High Temperature Gases." In: Trans. of Industrial Hygiene Foundation, 21st Annual Meeting Pittsburgh, Pa., 1956, p. 210–232. 15 refs.

Akerlow, E. V. "Modification to the Fontana Open Hearth Precipitators" J. Air Pollution Control Assoc. 7 (1), 39–43 (May 1957).

Brief, R. S., A. H. Rose, Jr., and D. G. Stephan. "Properties and Control of Electric-Arc Steel Furnace Fumes." J. Air Pollution Control Assoc. 6(4):220–204 (Feb. 1957).

Silverman, L. "High Temperature Gas and Aerosol Removal with Fibrous Filters." Proc. Air Water Pollution Abatement Conf., 1957. pp. 10–23m.

Spaite, P. W., D. G. Stephen, and A. H. Rose, Jr. "High Temperature Fabric Filtration of Industrial Gases." J. Air Pollution Control Assoc. 11, 243–7 & 58, May 1961.

Orban, A. R., J. D. Hummell, and G. G. Cocks. "Research on Control of Emissions from Bessemer Converters." J. Air Pollution Control Assoc. 11, (3) 103–13, Mar. 1961.

Harms, F. and W. Riemann. "Measurement of Fumes and Dust Volumes from 70-Ton Electric Arc Furnaces Operated Partially on Oxygen." Stahl Eisen, 82(20):1345–1348, 1962. Translated from German. Henry Brutcher Technical Translations, Altadena, Calif., HB-5719, 12 p., 1962.

Rengstorff, George W. P. "Factors Controlling Emissions from Steelmaking Processes." Open Hearth Proc., vol. 45:204–219, 1962. 7 refs.

Hunt, M. and A. T. Lawson. "The Control of Dust and Fume Emissions from an Integrated Steelworks." Proc. Clean Air Conf., Univ. New South Wales, 1962, Paper 15, Vol. 2, 33 p.

Campbell, W. W. and R. W. Fullerton. "Development of an Electric-Furnace Dust Control System." J. Air Pollution Control Assoc. 12 (12), 574–7; 590 (Dec. 1962).

Muhlrad, W. "Removal of Dust from Basic-Oxygen Furnace Brown Fumes by Means of Bag Filters." Stahl Eisen, 82(22):1579–1584, 1962. 9 refs. Translated from German. Henry Bruthcer, Technical Translations, Altadena, Calif., HB-5768, (29)p., 1963.

Schneider, R. L. "Engineering, Operation and Maintenance of Electrostatic Precipitators on Open Hearth Furnaces." J. Air Pollution Control Assoc. 13(8), 348–53 (Aug. 1963).

Pettit, Grant. A. "Electric Furnace Dust Control System." J. Air Pollution Control Assoc., 13(12):607–609, 621, Dec. 1963. 14 refs.

Punch, G. "Gas Cleaning in the Iron and Steel Industry. Part II: Applications." In: Fume Arrestment, Iron and Steel Inst., London (England), SR-83, p. 10–23, 1963. 34 refs.

Bintzer, W. W. "Design and Operation of a Fume and Dust Collection System for Two 100-Ton Electric Furnaces." Iron Steel Engr., 41(2):115–123, Feb. 1964.

Schueneman, J. J., M. D. High, and W. E. Bye. "Air Pollution Aspects of the Iron and Steel Industry." Public Health Service, Cincinnati, Ohio, Div. of Air Pollution. (999-AP-1) June 1963. 34 pp.

Benz, D. L. and R. Bird. "Control of Pollution from Electric Arc Furnaces at Oregon Steel Mills Co." Preprint. 1965.

Harris, E. R. and F. R. Beiser. "Cleaning Sinter Plant Gas with Venturi Scrubber." J. Air Pollution Control Assoc. 15, (2) 46–9, Feb. 1965.

Wheeler, D. H. and D. J. Pearse. "Fume Control Instrumentation in Steelmaking Processes." Blast Furnace Steel Plant 53 (12), 1125–30 (Dec. 1965.)

Eberhardt, J. E. and H. S. Graham. "The Venturi Washer for Blast Furnace Gas." Iron Steel Engr., 32(3):66–71, March 1965. 11 refs.

Nikami, K., K. Matsuda, T. Koyano, and T. Yasui. "The Waste Gases Leaving the Basic Oxygen LD Furnace." Tetsu To Hagane, 52(9):1491–1493, 1966. Translated from Japanese. Henry Brutcher, Technical Translations, Altadena, Calif., 8 p., 1967.

Broman, C. "Scrubbing for Clean Air." Preprint. (Presented at the 59th Annual Meeting, Air Pollution Control Association, San Francisco, Calif., June 20–24, 1966, Paper No. 66-99.)

Parker, C. M. "BOP Air Cleaning Experiences." Preprint. J. Air Pollution Control Assoc. 16, (8)446–8, Aug. 1966.

Hoak, R. D. and H. C. Brammer. "Pollution Control in the Steel Industry." Chem. Eng. Progr. 62, (10) 48–52, Oct. 1966.

Brandt, A. D. "Current Status and Future Prospects—Steel Industry Air Pollution Control." Proc. Natl. Conf. Air Pollution, 3rd, Washington, D.C., 1966. pp. 236–41.

Smith, W. M. and D. W. Coy. "Fume Collection in a Steel Plant." Chem. Eng. Progr., 62(7):119–123, July 1966.

Wilkinson, F. M. "Wet Washing of BOF Gases–Lackawanna." Preprint, Bethlehem Steel Corp., Lackawanna Plant, New York, 12 p., 1966.

Willett, H. P. "Cutting Air Pollution Control Costs." Chem. Eng. Progr. 63, (3) 80–3, Mar. 1967.

Johnson, J. E. "Wet Washing of Open Hearth Gases." Iron Steel Engr., 44(2): 96–98, Feb. 1967.

Bintzer, W. W. and D. R. Kleintop. "Design Operation and Maintenance of a 150-Ton Electric Furnace Dust Collection System." Iron Steel Engr. 44(1):77–85 June 1967.

Pottinger, J. F. "The Collection of Difficult Materials by Electrostatic Precipitation." Australian Chem. Process. Eng. (Sidney), 20(2):11–23, Feb. 1967. 11 refs.

Wilcox, Michael S. and Roy T. Lewis. "A New Approach to Pollution Control in an Electric Furnace Melt Shop." Iron Steel Engr., 45(12):113–120, Dec. 1968.

Kazarinoff, Andrew. "Industrial Air Pollution –Its Control and Cost." Design News, 23(14):18–24, July 5, 1968.

Sebesta, William. "Ferrous Metallurgical Processes." In: Air Pollution. Arthur C. Stern (ed.), Vol. 3, 2nd ed., New York, Academic Press, 1968, Chapt. 36, p. 143– 169. 40 refs.

Henschen, H. C. "Wet vs. Dry Gas Cleaning in the Steel Industry." J. Air Pollution Control Assoc., 18(5):338–342, May 1968.

Wheeler, D. H. "Fume Control in L-D Plants." J. Air Pollution Control Assoc., 18(2):98–101, Feb. 1968.

Hammond, William F., James T. Nance, and Karl D. Luedtke. "Steel-Manufacturing Processes." In: Air Pollution Engineering Manual. (Air Pollution Control District, County of Los Angeles.) John A. Danielson (comp. and ed.), Public Health Service, Cincinnati, Ohio, National Center for Air Pollution Control, PHS-Pub-999-AP-40, p. 141–257, 1967. GPO: 806-614-30.

Huntington, Robert G. and Donald H. Rullman. "Arc Furnace Fume Control Practices." Preprint, American Air Filter Co., Inc., Louisville, Ky., (19) p., 1968.

Venturini, J. L. "Historical Review of the Air Pollution Control Installation at Bethlehem Steel Corporation's Los Angeles Plant." Preprint. Bethlehem Steel Corp., Los Angeles, 22 p., 1968. 1 ref.

"A Status Report: Process Control Engineering; R & D for Air Pollution Control." Public Health Service, Cincinnati, Ohio, National Air Pollution Control Administration. 37 p., Nov. 1969.

Gedgaudas, Marius J. "The Emission Inventory and Its Application in the Iron and Steel Industry." Preprint. National Air Pollution Control Administration, Raleigh, N.C., Div. of Abatement, 13 p., April 1969.

Sullivan, Ralph J. "Preliminary Air Pollution Survey of Iron and Its Compounds. A Literature Review." Litton Systems, Inc., Silver Spring,

Md., Environmental Systems Div., NAPCA Contract PH 22-68-25, Pub. APTD 69-38, 94 p., Oct. 1969. 225 refs. CFSTI: PB 188088.

Varga, John, Jr. "A Systems Analysis Study of the Integrated Iron and Steel Industry. (Final Report)." Battelle Memorial Inst., Columbus, Ohio, NAPCA Contract PH 22-68-65, 518 p., May 15, 1969. 316 refs. NTIS: PB 184577.

Singhal, R. K. "Fume Cleaning Systems Used in the Steel Industry—Part I." Steel Times (London), 197(8):531–538, Aug. 1969.

Singhal, R. K. "Fume Cleaning Systems Used in the Steel Industry—Part II." Steel Times (London), 197(9):605–613, Sept. 1969.

Campbell, W. W. and R. W. Fullerton. "High-Energy Wet Scrubbers Can Satisfactorily Clean Blast Furnace Top Gas." American Inst. of Mining, Metallurgical and Petroleum Engineers (AIME), New York, N.Y., Proc. Am. Inst. Mining Mt. Petrol. Engrs. Conf. Blast Furnace, Coke Oven, Raw Materials 1969, vol. 18:329–335.

Squires, B. J. "Electric Arc Furnace Fume Control and Gas Cleaning." Conf. Filtration Soc., Dust Control Air Cleaning Exhibition, London, 1969, p. 16–21. 4 refs. (Sept. 23–25.)

"Air Pollution Aspects of Brass and Bronze Smelting and Refining Industry." U.S. Dept. of Health, Education & Welfare, Public Health Service, National Air Pollution Control Admin. Publication No. AP-58, Raleigh, N.C. Nov. 1969.

"National Inventory of Sources and Emissions. Cadmium, Nickel and Asbestos, 1968. Section I. Cadmium." Davis (W. E.) and Associates, Leawood, Kansas. NAPCA Contract CPA 22-69-NAPCA-APTD-68, 44 p., Feb. 1970. 12 refs. CFSTI:PB 192250.

Vandergrift, A. Eugene, Larry J. Shannon, Eugene E. Sallee, Paul G. Gorman, and William R. Park. "Particulate Air Pollution in the United States." Preprint, Air Pollution Control Assoc., Pittsburgh, Pa., 30 p., 1970. 2 refs.

Lownie, Harold W., Jr., and Thomas M. Barnes. "The NAPCA Study of Air Pollution in the Steel Industry." Preprint, Air Pollution Control Assoc., Pittsburgh, Pa., 40 p., 1970. 3 refs.

Giever, P. M. "Characteristics of Foundry Effluents." Preprint, American Foundrymen's Society, Des Plaines, Ill., 3 p., 1970. 7 refs.

Celenza, G. J. "Air Pollution Problems Faced by the Iron and Steel Industry." Plant Eng., 24(9):60–63, April 30, 1970.

Adams, Richard W. "High Energy Wet Gas Cleaning for the Basic Oxygen OG Process." Blast Furn. Steel Plant, 58(10):751–753, Oct. 1970.

Egley, Billy D. "Selection of Gas Cleaning Equipment for an Ore Preparation Plant." Iron Steel Engr., 47(11):111–115, Nov. 1970.

Venturini, J. L. "Operating Experiences with a Large Baghouse in an Electric Arc Furnace Steelmaking Shop." J. Air Pollution Control Assoc., 20(12): 808–813, Dec. 1970. 2 refs.

Talbott, John A. "Building a Pollution-Free Steel Plant." Mech. Eng., 93(1): 25–30, Jan. 1971.

Richard, Jablin. "Environmental Control at Alan Wood: Technical Problems, Regulations and New Processes." Iron Steel Engr., 48(7):58–65, July 1971. 7 refs.

"Problems of Dust, Fume, and Effluent Control in the Iron and Steel Industry." J. Iron Steel Inst., vol. 209:25–26, 28–30, 32, 34, 36–43, June 1971.

Elliott, A. C. and A. J. Lafreniere. "The Design and Operation of a Wet Electrostatic Precipitator to Control Billet Scarfing Emissions." Preprint, Air Pollution Control Assoc., Pittsburgh, Pa., 10 p., 1971.

Bramer, Henry C. "Pollution Control in the Steel Industry." Environ. Sci. Technol., 5(10):1004–1008, Oct. 1971. 6 refs.

Brough, John R. and William A. Carter. "Air Pollution Control of an Electric Furnace Steelmaking Shop." Preprint, Air Pollution Control Assoc., Pittsburgh, Pa., 31 p., 1971. 4 refs.

GENERAL REFERENCES

Section IV

Plass, Robert J. and Harold H. Haaland. "Electrostatic Precipitators in the Cement Industry." J. Air Pollution Control Assoc., 9(2):96–97, 100–101, Aug. 1959, 6 refs.

O'Mara, Richard F. and Carl R. Flodin. "Filters and Filter Media for the Cement Industry." J. Air Pollution Control Assoc., 9(2):96–97, 100–101, Aug. 1959. 6 refs.

White, Harry J. and Walter A. Baxter, Jr. "A Superior Collecting Plate for Electrostatic Precipitators." Preprint, American Society of Mechanical Engineers, New York, 7 p., 1959. 2 refs.

"Air-Borne Particulate Emissions from Cotton Ginning Operations." United States Dept. of Health, Education and Welfare, Cincinnati, Ohio, U.S. Public Health Service Tech. Report A 60-5, 1960.

"Guide for Air Pollution Control of Hot Mix Asphalt Plants." National Asphalt Pavement Assoc., Information Series 17, 1965.

Tomaides, M. "Dust Collection in the Cement Industry." Proc. (Part I) Intern. Clean Air Cong., London, 1966. (Paper V/4). pp. 125–8.

Doherty, R. E. "Current Status and Future Prospects–Cement Mill Air Pollution Control." Proc. Natl. Conf. Air Pollution 3rd, Washington, D.C. 1966. pp. 242–9.

Hankin, Montagu, Jr. "Various Methods of Dust Collection at Stone Plants." Preprint, Grove (N.J.) Lime Co., Lime Kiln, Md., 35 p., 1966. 5 refs.

"Restriction of Emission Portland Cement Works." VDZ (Verein Deutscher Zementwerke), Emissionsausschuss. (Auswurfbegrenzung Zementwerke). VDI (Ver. Deut. Ingr.) Richtlinien, no. 2094, Feb. 1967. 15 refs. Translated from German by D. Ben Yaakov, Israel Program for Scientific Translations, Jerusalem, 26 p. CFSTI:TT 68-50469/12.

Hankin, M., Jr. "Is Dust the Stone Industry's Next Major Problem?" Rock Prod., 70(4):80–4, 110, Apr. 1967.

Kreichelt, T. E., D. A. Kemnitz, S. T. Cuffe. "Atmospheric Emissions from the Manufacture of Portland Cement." Public Health Service, Cincinnati, Ohio, National Center for Air Pollution Control, PHS-Pub-999-AP-17, 47 p., 1967. 29 refs. GPO:803-789-2.

Wiemar, Peter. "Dust Removal from the Waste Gases of Preparation Plants for Bituminous Road-Building Materials." Staub (English translation), 27(7):9–22, July 1967. 2 refs. CFSTI:TT 67-51408/7.

Vincent, Edwin J. and John L. McGinnity. "Concrete-Batching Plants." In: Air Pollution Engineering Manual. (Air Pollution Control District,

County of Los Angeles.) John A. Danielson (comp. and ed.), Public Health Service, Cincinnati, Ohio, National Center for Air Pollution Control, PHS-Pub-999-AP-40, p. 334–339, 1967. GPO: 806-614-30.

Vincent, Edwin J. "Rock and Gravel Aggregate Plants." In: Air Pollution Engineering Manual. (Air Pollution Control District, County of Los Angeles.) John A. Danielson (comp. and ed.), Public Health Service, Cincinnati, Ohio, National Center for Air Pollution Control, PHS-Pub-999-AP-40, p. 340–342, 1967. GPO: 806-614-30.

"Control and Disposal of Cotton-Ginning Wastes." U.S. Dept. of HEW. A Symposium Sponsored by National Center for Air Pollution Control, Public Health Service and Agricultural Engineering Research Div., U.S. Dept. of Agriculture, May 3–4, 1966, Dallas, Texas, Public Health Service Pub. No. 999-AP-31, 1967.

Muhlrad, M. Wolf. "Cement Plants and Atmospheric Pollution. Problems of Dust Removal." (Les cimenteries et la pollution atmospherique. Les problemes de depoussierage). Text in French. Equipement Mecan., 48(87): 91–95, 1969.

Herod, Buren C. "NCSA's Dust Control Seminars Reflect Industry's Concern with Effective Measures." Pit and Quarry, 61(12):118–124, June 1969.

Tripler, Arch B., Jr., and G. Ray Smithson, Jr. "A Review of Air Pollution Problems and Control in the Ceramic Industries." Preprint, American Ceramic Society, Columbus, Ohio, 25 p., May 5, 1970. 19 refs.

Trauffer, Walter E. "Maine's New Dust-Free Crushed Stone Plant." Pit Quarry, 63(2):96–100, Aug. 1970.

Lyn, Andrew Van der. "Prescription for Cement Plant Dust Control." Rock Prod., vol. 73:73, 76–78, 80, 86, 87, Aug. 1970.

Lyn, Andrew Van der. "Prescription for Cement Plant Dust Control. Part 2." Rock Prod., 73(9):118–120, 136–138, Sept. 1970. Part 1. Ibid, Aug. 1970.

"Environmental Pollution Control at Hot Mix Asphalt Plants." National Asphalt Pavement Assoc., Information Series 27, Riverdale, Maryland, 1970.

GENERAL REFERENCES

Section V

Clauss, N. W. "The Reduction of Atmospheric Pollution from Sulfuric and Recovery Processes." Manufacturing Chemists Association, Washington, D.C. Air Pollution Abatement Committee and Manufacturing Chemists Association, Washington, D. C., Water Pollution Abatement Committee Proc. Mfg. Chem. Ass. 1952–53 Pollution Abatement Conference, 9 p.

Hitchcock, L. B. "Air Pollution and the Oil Industry." Proc. Am. Petrol. Inst., Sect. IV. 35, 150–4, 1955.

Wright, R. H. "Is It Possible to Build and Operate a Completely Odorless Kraft Mill?" Can. Pulp Paper Ind. (Vancouver), 10(9):21–22, 24, 26, 28, 30, 32, 34, Sept. 1957.

Boubel, R. W., M. Northcraft, A. Van Vllet, M. Papovich. "Wood Waste Disposal and Utilization." Engineering Experiment Station Bulletin No. 39, Oregon State College, Corvallis, Oregon, August 1958.

Steigerwald, Bernard J. and A. H. Rose. "Atmospheric Emissions from Petroleum Refineries. A Guide for Measurement and Control." Public Health Service, Cincinnati, Ohio Div. of Air Pollution, PHS Pub.-763, 56 p., 1960. 11 refs. NTIS: PB 198096.

"Sulfuric Acid Manufacture, Report No. 2." J. Air Pollution Control Assoc. 13, (10)499-502, Oct. 1963.

Avy, A. P. "Methods of Reducing Pollution Caused by Specific Industries. (Chapter VI. Chemical Industry)." European Conf. of Air Pollution, Strasburg, 1964. p. 337–356.

Cuffe, S. T. and C. M. Dean. "Atmospheric Emissions from Sulfuric Acid Manufacturing Process; A Comprehensive Abstract." Preprint. (Presented at the 58th Annual Meeting, Air Pollution Control Association, Toronto, Ontario, Canada, June 1965.)

"Air Pollution Control, National Fertilizer Development Center, Wilson Dam, Alabama." Tenn. Ind. Hyg. News., 22(1):1–5, Winter 1965. APTIC No. 32231.

Boubel, R. W. "Wood Residue Incineration in Tepee Burners." Engr. Experiment Station Circular No. 34, Oregon State College, Corvallis, Oregon, July 1965.

Gartrell, F. E. and J. C. Barber. "Pollution Control Interrelationships." Chem. Eng. Prog., 62(10):44–47, Oct. 1966, 3 refs. APTIC No. 32232.

Termeulen, M. A. "Air Pollution Control by Oil Refineries." Proc. (Part I) Intern. Clean Air Cong., London, 1966 (Paper IV/5). pp. 92–5.

334   GENERAL REFERENCES

Gammelgard, P. N. "Current Status and Future Prospects–Refinery Air Pollution Control." Preprint, 13 p., (1966). (Presented at the National Conference on Air Pollution, Washington, D.C., Dec. 13, 1966.)

Gerstle, R. W. and R. F. Peterson. "Atmospheric Emissions from Nitric Acid Manufacturing Processes–A Comprehensive Summary." Preprint. (For Presentation at the American Inst. of Chemical Engineers, Detroit, Mich., Dec. 8, 1966.)

Sundaresan, B. B., C. I. Harding, F. P. May, and E. R. Hendrickson. "A Dry Process for the Removal of Nitrogen Oxides from Waste Gas Streams in Nitric Acid Manufacture." Preprint. (Presented at the 59th Annual Meeting, Air Pollution Control Association, San Francisco, Calif., June 20–25, 1966, Paper 66-96.)

"Atmospheric Emissions from Nitric Acid Manufacturing Processes." Public Health Service, Cincinnati, Ohio, Div. of Air Pollution and Manufacturing Chemists Association, Washington, D.C. 1966. 96 pp. (999-AP-27.)

Anderson, H. C., P. L. Romeo, and W. J. Green. "A New Family of Catalysts for Nitric Acid Tail Gases." Engelhard Ind. Tech. Bull. 7(3), 100-5 (Dec. 1966).

Hendrickson, E. R., ed. "Atmospheric Emissions from Sulfate Pulping." (Proceedings of the International Conference on Atmospheric Emissions from Sulfate Pulping, April 28, 1966, Sanibel Island).

Anderson, H. C., P. L. Romeo, and W. J. Green. "A New Family of Catalysts for Nitric Acid Tail Gases." Nitrogen, no. 50:33–36, Nov./Dec. 1967. 6 refs.

Dickey, S. W. and C. W. Phillips. "Air Pollution Control Features of a Modern Refinery." American Chemical Society, Div. of Petroleum Chemistry Inc. and American Chemical Society, Pittsburgh, Pa., Div. of Water, Air and Waste Chemistry, American Chemical and Society Joint Symposium on Experience with Pollution Control Equipment, Chicago, Ill., 1967, p. A41–A42. (Sept. 11–15.)

Ross, L. W. and H. C. Lewis. "The Reaction of Sulfur Oxides with Phosphate Rock." Ind. Eng. Chem., 6(4):407–408, Oct. 1967. 10 refs.

Douglass, Irwin B. and Lawrence Price. "Sources of Odor in the Kraft Process. II. Reactions Forming Hydrogen Sulfide in the Recovery Furnace." TAPPI, 51(10):465–467, Oct. 1968. 9 refs. APTIC No. 28885.

Sheppard, Stanton V. "Control of Noxious Gaseous Emissions." American Inst. of Chemical Engineers, New York, American Inst. of Mining, Metallurgical and Petroleum Engineers (AIME), New York, N.Y., American Society of Civil Engineers, New York, American Society of Heating, Refrigerating and Air-Conditioning Engineers, New York, American Society of Mechanical Engineers, New York, and American Society for Testing and Materials, New York, Engineering Foundation, Proc.

MECAR Symp. New Developments Air Pollution Control, New York, 1967, p. 21–29. (Oct. 23.)

"Report of Investigation on the Effect of Waste Gases etc. from Manufacturing Factory of Vinyl Chloride Product." Japan Environmental Sanitation Association, Kawasaki. (Enka-biniru seihin kakoseizo kojo kara no haigasuto eikyo chosa hokokusho). Text in Japanese. 20 p., Jan. 1968.

Russell, W. E. "The Recovery of Fluoride from Superphosphate Manufacture." Chem. Ind. New Zealand, 4(11):10–11, 13, Nov. 1968. 11 refs.

Karbe, K. "Fluoride Emission from Fertilizer Production." Chem. Engr. (London), 46(7):CE268, Sept. 1968.

Heller, Austin N., Stanley T. Cuffe, and Don R. Goodwin. "Inorganic Chemical Industry." In: Air Pollution. Arthur C. Stern (ed.), Vol. 3, 2nd ed., New York, Academic Press, 1968, Chapt. 38, p. 191–242, 84 refs.

Elkin, Harold F. "Petroleum Refinery Emissions." In: Air Pollution. Arthur C. Stern (ed.), Vol. 3, 2nd ed., New York, Academic Press, 1968, Chapt. 34, p. 97–121. 23 refs.

Bingham, E. C. Jr. "Nitric Acid Plant's Catalytic Burner Prevents Tail Gas Stream Air Pollution." Preprint, Farmers Chemical Association, Inc., Tyner, Tennessee, 3 p., 1968.

Teske, W. "Emissions and Abatement of Oxides of Nitrogen in Nitric Acid Manufacture." Chem. Eng., No. 221, CE263–266, Sept. 1968.

Boubel, R. W. "Particulate Emissions from Sawmill Waste Burners." Engr. Experiment Station Bulletin No. 42, Oregon State University, Corvallis, Oregon. Aug., 1968.

Weiner, Jack and Lillian Roth. "Air Pollution in the Pulp and Paper Industry." Inst. Paper Chem., Bibliog. Ser., no. 237:1–224, 1969. 769 refs.

Hasegawa, T., H. Yoshikawa, S. Iwasaki, K. Shinra, and T. Shono. "Air Pollutant Emissions from Plastics Manufactories." (Gohsei jushi kakoh kohjoh no hai gas yugai seiboon chohsa [Dai 2 hoh]). Text in Japanese. Taiki Osen Kenkyo (J. Japan Soc. Air Pollution), 4(1):84, 1969.

"Air Pollution." Chem. Process., 16(11):11, 13, Dec. 1970. 7 refs.

Ermenc, E. D. "Controlling Nitric Oxide Emission." Chem. Eng., 77(12):193–196, June 1, 1970. 1 ref.

Donovan, J. R. and P. J. Stuber. "Air Pollution Shashed at Sulfuric-Acid Plant." Chem. Eng., 77(25): 47–49, Nov. 30, 1970.

Triplett, Gary. "Estimation of Plant Emissions." Preprint, p. 15–27. 1970. 21 refs.

Rasmussen, R. A. "Qualitative Analyses of the Hydrocarbon Emission from Veneer Dryers." Washington State University, Pullman, Air Pollution Research Section, Air Pollution Control Office, Research Grant AP 1232, 17 p., 1970.

Rathgeber, Ferdinand. "Dust and Shaving Removal in Woodworking Plants." (Beseitigung von Staub und Spaenen in holzverarbeitenden Betrieben). Text in German. Wasser Luft Betrieb, 14(8):312–318, Aug. 1970.

"Atmospheric Emissions from Wet-Process Phosphoric Acid Manufacture." NAPCA Pub. AP-57, 86 p., April 1970. 39 refs. CFSTI:PB 192222.

Wolpert, V. "Pollution Control in the Polymer Industries." Polym. Age, 1(6):260–261, Nov. 1970.

Hendrickson, E. R., J. E. Roberson, and J. B. Koogler. "Control of Atmospheric Emissions in the Wood Pulping Industry. Volume 3. (Final Report)." Environmental Engineering, Inc., Gainesville, Florida, and Stirrine (J. E.) Co., Greenville, S.C., Contract CPA 22-69-18, 250 p., March 15, 1970. 418 refs. CFSTI:PB 190353.

Hardison, L. C. "Techniques for Controlling the Oxides of Nitrogen." J. Air Pollution Control Assoc., 20(6):377–382, June 1970. 10 refs.

Alferova, L. A. and G. A. Titova. "Oxidation of Sodium Sulfide and Mercaptide in Black Liquor." Bumazhn. Prom. (Moscow), 41(10):5–6, Oct. 1966. Translated from Russian by Brenda Jacobsen, Washington Univ., Seattle, Dept. of Civil Engineering, 11 p., Oct. 31, 1970.

Kosaya, G. S. "Oxidation of Black Liquor with Oxygen." Bumazhn. Prom. (Moscow), 31(6):15, June 1956. 4 refs. Translated from Russian by Brenda Jacobsen, Washington Univ., Seattle, Dept. of Civil Engineering, 5 p., Sept. 28, 1970.

Shigeta, Yoshihiro. "Odor Pollution in the Chemical Plant and Its Measurement Method." (Kagaku kojo ni okeru akushukogai to snon sokuteiho). Text in Japanese. Anzen Kogaku (J. Japan Soc. Safety Eng.), 9(1):20–28, Feb. 1970. 19 refs.

Blosser, Russell O., Andre L. Caron, and Leon Duncan. "An Inventory of Miscellaneous Sources of Reduced Sulfur Emissions from the Kraft Pulping Process." Preprint, Air Pollution Control Assoc., Pittsburgh, Pa., 13 p., 1970. 2 refs.

Davis, John C. "Pulpers Apply Odor Control." Chem. Eng., 78(13):52–54, June 14, 1971.

"One Answer to Plasticizer Pollution." Mod. Plastics, 48(6):48–49, June 1971.

Mueller, James H. "What It Costs to Control Process Odors." Food Eng., 43(4):62–65, April 1971.

Roberson, James E., E. R. Hendrickson, and W. Gene Tucker. "The NAPCA Study of the Control of Atmospheric Emissions in the Wood Pulping Industry." TAPPI, 52(2):239–244, Feb. 1971.

Popovici, N., P. Potop, L. Brindus, and S. Nicolescu. "Process and Plant to Render Valuable SO_2 and NH_3 Waste Gaseous Components Produced by a Complex Fertilizer Plant." Chem. Oil Gas Rom., 7(1):21–29, 1971. 1 ref.

Waid, Donald E. "The Control of Odors by Direct Fired Gas Thermal Incineration." Preprint, Air Pollution Control Assoc., Pittsburgh, Pa., 50 p., 1971.

"Environmental Protection Problems of the Wood Processing Industry to be Analyzed." (Skegsindustrins miljovardsfragor utreds). Text in Swedish. Tek. Tidskr., 101(1):54–55, Jan. 14, 1971.

Maier, Alfred. "Protection Against Emission in the Woodworking Industry." (Immissionsschutz beim holzbearbeitenden und -verarbeitenden Gewerbe). Text in German. Wasser Luft Betrieb, 15(6):214–219, June 1971. 8 refs.

Maier, Alfred. "Emmission Protection in the Wood Working Industry." (Immissionsschutz beim holzbearbeitenden and und -verarbeitenden Gewerbe). Text in German. Wasser Luft Betrieb, 15(7):261–264, July 1971.

Ricketts, C. "The Control of Offensive Odors: Results of a Survey." Environ. Health, 79(5):136–138, 140–142, May 1971.

Wrist, P. E. "Impact of New Air Pollution Regulations on the Pulp and Paper Industry." Tappi, 54(7):1090–1093, July 1971.

Sherwood, R. J. "Trends in the Refinery Environment." Med. Bull. Standard Oil, New Jersey, 31(2):142–156, July 1971. (Also: Petrol. Rev., Feb. 1971.)

INDEX